Biocarbon Polymer Composites

Edited by

Deepa Kodali

Department of Mechanical Engineering
Christian Brothers University, Memphis, TN 38104
USA

&

Vijaya Rangari

Department of Materials Science and Engineering
Tuskegee University, Tuskegee, AL 36088
USA

Biocarbon Polymer Composites

Editors: Deepa Kodali & Vijaya Rangari

ISBN (Online): 978-981-5196-68-9

ISBN (Print): 978-981-5196-69-6

ISBN (Paperback): 978-981-5196-70-2

© 2023, Bentham Books imprint.

Published by Bentham Science Publishers Pte. Ltd. Singapore. All Rights Reserved.

First published in 2023.

need for a court order if at any point you breach any terms of this License Agreement. In no event will any delay or failure by Bentham Science Publishers in enforcing your compliance with this License Agreement constitute a waiver of any of its rights.

3. You acknowledge that you have read this License Agreement, and agree to be bound by its terms and conditions. To the extent that any other terms and conditions presented on any website of Bentham Science Publishers conflict with, or are inconsistent with, the terms and conditions set out in this License Agreement, you acknowledge that the terms and conditions set out in this License Agreement shall prevail.

Bentham Science Publishers Pte. Ltd.
80 Robinson Road #02-00
Singapore 068898
Singapore
Email: subscriptions@benthamscience.net

BENTHAM SCIENCE

CONTENTS

PREFACE

The editors are pleased and excited to present the book titled ***Biocarbon Polymer Composites***. The title of the book reflects the authors' awareness of the significance of biocarbon-based composites in the modern industrial and manufacturing fields, as well as their desire to introduce readers to one of the most unique classes of materials.

The 21ˢᵗ century has seen an increasing global demand for materials that are sustainable, ecofriendly, and biodegradable. The growing concerns about the environmental impact of plastic waste have spurred research and development of alternative materials, such as biocarbon polymer composites. Dr. Jane Goodall, the internationally acclaimed environmentalist said that *Biocarbon polymer composites have the potential to revolutionize the way we manufacture products, by providing an environmentally friendly alternative to traditional plastics*. As said, one of the remarkable features of biocarbon polymer composites is that they have the potential to reduce our dependence on non-renewable resources, mitigate the environmental impact of plastic waste, and create new economic opportunities for rural communities. Biocarbon polymer composites are an exciting new option as we seek out greener alternatives to petroleum-based plastics.

The book is organized into three sections and has nine chapters in total. The three sections Section I - State of the Art, Section II is the Sources of biocarbon and preparation of polymer composites, and Section III is the Properties and Applications of biocarbon polymer composites. Section I includes Chapters 1 and 2, Section II contains Chapter 3, and Section III contains Chapters 4 through 9. Chapter 1 introduces the use of bio nanocarbon to enhance fiber-reinforced nano biocomposites, addressing challenges in composite properties. It covers biocarbon nanoparticle synthesis and their integration into polymeric composites, with a focus on bio-inspired structures. Chapter 2 explores the circular economy and biochar production from waste, emphasizing its potential as a filler for polymer matrices. Section II, Chapter 3, discusses the use of animal waste-derived biochar in polymer nanocomposites, highlighting improved mechanical and thermal properties. Chapter 4 in Section III delves into biochar's application in thermal energy storage materials. Chapter 5 explores biocarbon extraction from various natural waste sources for 3D printing. Chapter 6 investigates rice husk ash as a sustainable filler for biocomposites, analyzing its impact on tensile properties. Chapter 7 discusses biochar-based nanocomposites and their enhanced electrochemical properties, particularly in agriculture. Chapter 8 focuses on synthesizing high-quality graphitic carbon from coconut shell powder for LDPE composite films with improved properties. Lastly, Chapter 9 emphasizes the environmental importance of polymer nanocomposites and biocarbon-reinforced composites in pollution remediation and sustainability efforts.

The main aim of this book is to offer a thorough introduction to biocarbon polymer composites, including the most recent research and practical applications of this material. This book brings together contributions from prominent researchers, engineers, and industry professionals in the field of biocarbon polymer composites. The chapters are structured to provide a clear and concise understanding of biocarbon-based composites' fundamental concepts and practical applications. This book is a must-read for anyone interested in the science of these materials and their prospective uses.

Deepa Kodali
Department of Mechanical Engineering
Christian Brothers University, Memphis, TN 38104
USA

&

Vijaya Rangari
Department of Materials Science and Engineering
Tuskegee University, Tuskegee, AL 36088
USA

ACKNOWLEDGEMENT

First and foremost, we would like to express our sincere gratitude to God, the almighty, with whose grace that we are able to complete this book. His blessings have given us the strength and perseverance to pursue our passion and bring this project to fruition. We are also deeply grateful to our parents and families for their unwavering support and encouragement throughout this journey. Their patience, understanding, and love have been constant source of inspiration and motivation. Their belief in us and our work kept us motivated and helped us overcome any obstacles that came our way. We are deeply indebted to our friends for their insightful feedback, constructive criticism, and stimulating discussions. Their contributions have helped shape the ideas and refine the writing, and we are truly grateful for their friendship and camaraderie. We would also like to extend our thanks to all the contributing authors who helped make this book possible. Their expertise and knowledge have greatly enriched this work, and we are grateful for their contribution.

We extend our heartfelt gratitude to the contributing authors, reviewers, editors, advisory board members and the team of Bentham Books for their invaluable guidance and support throughout the writing process. We are immensely grateful for their contributions and for believing in our vision for this project. We would like to express our sincere appreciation to everyone who has played a role in making this book a reality, from those who provided encouragement and support throughout the writing process to those who provided technical assistance in formatting and publishing. Your help has been essential, and we are truly grateful for your contributions.

Finally, our heartfelt thanks to all the students, researchers, and professionals who will read this book. We hope that this book will be an invaluable resource and inspire and inform the next generation of scientists and engineers who will shape the future of materials science. We believe that these materials represent a crucial step forward in the development of sustainable and eco-friendly alternatives to standard plastics, and we are delighted to contribute to this important and timely topic with this book.

Deepa Kodali
Department of Mechanical Engineering
Christian Brothers University, Memphis, TN 38104
USA

&

Vijaya Rangari
Department of Materials Science and Engineering
Tuskegee University, Tuskegee, AL 36088
USA

List of Contributors

Aman Sharma   Department of Chemistry, CHRIST (Deemed to be University), Hosur Road, Bengaluru 560029, India
Centre for Advanced Research and Development (CARD), CHRIST (Deemed to be University), Hosur Road, Bengaluru 560029, India

B. Anjaneya Prasad   Department of Mechanical Engineering, Jawaharlal Nehru Technological University, Hyderabad, Telangana 500085, India

Esperidiana A.B. Moura   Centro de Química e Meio Ambiente, Instituto de Pesquisas Energéticas e Nucleares, Av. Prof. Lineu Prestes, 2242, São Paulo 05508-000, SP, Brazil

Fernando L. Almeida   Faculdade de Tecnologia do Estado de São Paulo "Prof. Miguel Reale", Av. Miguel Ignácio Curi, 360, São Paulo 08295-005, SP, Brazil

Gautam Chandrasekhar   Department of Materials Science and Engineering, Tuskegee University, Tuskegee, AL 36088, USA

Giulio Malucelli   Department of Applied Science and Technology, Politecnico di Torino, Viale Teresa Michel 5, 15129, Alessandria, Italy
Consorzio Interuniversitario Nazionale per la Scienza e Tecnologia dei Materiali (INSTM),Via G. Giusti 9, 50121, Firenze, Italy

Gowri Soman   Department of Chemistry, CHRIST (Deemed to be University), Hosur Road, Bengaluru 560029, India
Centre for Advanced Research and Development (CARD), CHRIST (Deemed to be University), Hosur Road, Bengaluru 560029, India

Gurumurthy Hegde   Department of Chemistry, CHRIST (Deemed to be University), Hosur Road, Bengaluru 560029, India
Centre for Advanced Research and Development (CARD), CHRIST (Deemed to be University), Hosur Road, Bengaluru 560029, India

Gustavo F. Souza   Centro de Química e Meio Ambiente, Instituto de Pesquisas Energéticas e Nucleares, Av. Prof. Lineu Prestes, 2242, São Paulo 05508-000, SP, Brazil

Janetty J.P. Barros   Centro de Química e Meio Ambiente, Instituto de Pesquisas Energéticas e Nucleares, Av. Prof. Lineu Prestes, 2242, São Paulo 05508-000, SP, Brazil
Academic Unit of Materials Engineering, Federal University of Campina Grande, Campina Grande 58249-140, Brazil

Jasmine Joseph   Department of Chemistry, CHRIST (Deemed to be University), Hosur Road, Bengaluru 560029, India
Centre for Advanced Research and Development (CARD), CHRIST (Deemed to be University), Hosur Road, Bengaluru 560029, India

Joao A. Antonangelo   Washington State University , Pullman, WA, USA

Kiran Bijapur   Department of Chemistry, CHRIST (Deemed to be University), Hosur Road, Bengaluru 560029, India
Centre for Advanced Research and Development (CARD), CHRIST (Deemed to be University), Hosur Road, Bengaluru 560029, India

Mahesh K. Pallikonda   Department of Engineering Technology, Austin Peay State University, Clarksville, TN 38104, USA

Niranjan Thir Department of Mechanical Engineering, Mahatma Gandhi Institute of Technology, Hyderabad, Telangana-500075, India

Radhika Mandala Department of Mechanical Engineering, Jawaharlal Nehru Technological University, Hyderabad, Telangana 500085, India
Department of Mechanical Engineering, Vignan Institute of Technology and Science, Deshmukhi, Hyderabad, Telangana 508284, India

Rene R. Oliveira Centro de Ciências e Tecnologia de Materiais, Instituto de Pesquisas Energéticas e Nucleares, Av. Prof. Lineu Prestes, 2242, São Paulo 05508-000, SP, Brazil

Rutika Umesh Kankrej Department of Mechanical Engineering, Vignan Institute of Technology and Science, Hyderabad, Telangana 508284, India

Sabina Yeasmin Department of Microbiology, CNMC, Kolkata, West Bengal, India

Singaravel Balasubramaniyan Department of Mechanical Engineering, Vignan Institute of Technology and Science, Hyderabad, Telangana 508284, India

Soma Bose Department of Microbiology, CNMC, Kolkata, West Bengal, India

Suresh Akella Department of Mechanical Engineering, Sreyas Institute of Engineering and Technology, Naderagul, Hyderabad, Telangana 500068, India

Vandana Molahalli Department of Chemistry, CHRIST (Deemed to be University), Hosur Road, Bengaluru 560029, India
Centre for Advanced Research and Development (CARD), CHRIST (Deemed to be University), Hosur Road, Bengaluru 560029, India

Venkateswara Rao Kode Department of Chemical and Biochemical Engineering, Christian Brothers University, Memphis, TN 38104, USA

Vijaya Rangari Department of Materials Science and Engineering, Tuskegee University, Tuskegee, AL 36088, USA

CHAPTER 1

Synergistic Effect of Bio-Nanocarbon Embedded Polymer Nanocomposite and its Applications

Vandana Molahalli[1,2]**, Jasmine Joseph**[1,2]**, Kiran Bijapur**[1,2]**, Aman Sharma**[1,2]**, Gowri Soman**[1,2] **and Gurumurthy Hegde**[1,2,*]

[1] *Department of Chemistry, CHRIST (Deemed to be University), Hosur Road, Bengaluru 560029, India*

[2] *Centre for Advanced Research and Development (CARD), CHRIST (Deemed to be University), Hosur Road, Bengaluru 560029, India*

Abstract: For applications involving sustainable materials, bio-nanocarbon was examined as a material to improve the properties of fiber-reinforced nano-biocomposite. A thorough investigation has been conducted using nano biocarbon as a filler and reinforcing material. However, the composite's inferior mechanical, physical, and thermal properties are a result of a poor fiber-matrix interface. As a result, in this study, biocarbon nanoparticles were created and used as functional components to enhance the properties of polymeric composite materials. To emphasize the scientific and technological issues that need to be resolved in order to create artificial composites with bio-inspired structures, recent studies of bio-inspired nano-carbon composites are discussed in this study. These include the production techniques for resolving the nano-carbon dispersion problem and creating bio-inspired structures, as well as the microstructure and composite characteristics characterization. In order to reveal natural design principles and serve as a resource for future research, bio-inspired composites and their applications are thoroughly examined and explained.

Keywords: Biocarbon, biochar, nano-biocomposite, nano-carbon, microstructure.

1. INTRODUCTION

Carbon is one of the periodic table's most versatile elements, having unrivaled qualities [1]. The atomic organization and hybridization of carbon determine its properties as an insulator (diamond) or a semiconductor (stacked graphite) or, amorphous carbon in a high-floor location. The ability of carbon to catenate opens the door for expanding the various fields of science such as chemistry, physics,

---

* **Corresponding author Gurumurthy Hegde:** Department of Chemistry, CHRIST (Deemed to be University), Hosur Road, Bengaluru 560029, India; E-mail: murthyhegde@gmail.com

Deepa Kodali & Vijaya Rangari (Eds.)

and biology to facilitate people's lives. Carbon nanostructures, including activated carbon (AC), carbon nanotubes (CNT), graphite, fullerene, carbon quantum dots, *etc.*, are low dimensional allotropes of carbon emerging as bright spots in the domains of science and technology. Carbon allotropes are the primary building blocks of biological life. Carbon is used by nature together with other elements to generate a variety of species and offers sustainable ways to change matter and energy. These carbon-based resources, or biomass, have been used as renewable starting materials for the controlled manufacture of carbon compounds, which might minimize the usage of fossil-based fuels and hasten the sustainable growth of human society. Carbon is found in all living organisms, including plants, animals, and humans. So, nature is the best source for deriving carbon from agricultural waste, forest, and household waste. Synthesizing carbon nanomaterials from biomass is cost-effective and environmentally friendly. Many studies have been conducted to transform carbon from bio-resources to nanomaterials for diverse uses. Sugarcane bagasse, bamboo, paper pulp sludge, fruit and vegetable peels, rice husk, oil seeds, coconut shells, and other bioresources have been used as carbon precursors. Lignocellulose in biomass is the primary precursor for synthesizing biocarbon nanomaterials [2]. Biomass can be either lignocellulosic or non-lignocellulosic depending on the composition. Most biomass is lignocellulosic. Cellulose, hemicellulose, and lignin are the three distinct constituents of non-edible biomass generated from agriculture and forest wastes. Carbonaceous materials can also be made from non-lignocellulosic biomass including fruit, animal, and food waste, which is high in carbohydrates, polysaccharides, and protein. It is important to note that the heterogeneous chemical components and structure of biomass make the production of homogenous and controlled carbonaceous materials challenging [3]. The BCMs (Biocarbon materials) made from plant- and animal-derived biomass often had some noticeable variations, particularly in their structural and chemical make-ups. For instance, doping-free wood-based BCMs mostly include C and O and may readily maintain the original biomass structures. The crucial step in synthesizing BCMs with the necessary shapes and morphologies is selecting the appropriate biomass precursors. The majority of biomass resources are made up of complex compounds, and these molecules often have various pyrolysis pathways [4]. BCM topologies and architectures are important factors in determining applications. It depends on whether the BCMs have inherited the original macro- or micro-structures of the biomass. Numerous compounds found in biomass vary in kind and amount from one species to another, and as a result, have unique breakdown pathways. In general, when plant biomass is transformed into BCMs *via* an hydrothermal carbonization method, the original structure may be preserved since it comprises more crystalline cellulose and less hemicellulose. The majority of chitin biomass may be immediately pyrolyzed into BCMs. Proteins have been routinely employed to create BCMs that include nitrogen since they are necessary for all living things. The ultimate architectures and characteristics of the BCMs can be influenced by a variety of circumstances [5].

Thermochemical conversion of biomass resulted in the production of biochar and carbon fibers initially. Later on, the possibility of synthesizing carbon nanomaterials from biomass expanded the world of biomass as nanocarbon precursors. CNT, fullerene, quantum dots, graphite and graphene, and carbon nanosphere are also prepared from biomass precursors [5]. Biomass feedstocks go through a sequence of simultaneous events during pyrolysis, including dehydration, decarboxylation, aromatization, and recondensation.

The surface morphology and porous nature of carbon materials were investigated using FE-SEM images (Fig. **1**). The formation of thick and large-sized carbon particles with a smooth surface could be seen in the images. This clearly confirms that the carbonization process at different temperatures from 600 to 800 °C in an Ar or N_2 atmosphere could convert the dried biowaste into a valuable carbon product. However, all raw samples were characterized by a dense structure. After the pyrolysis processes, the pores were created. In case of bio-chars from wheat straw, a small amount of pores was observed on the surface. Their amount was increased with increasing the process temperature, but the pores were very small. The SEM images of the all obtained activated carbons show more porous structures. For all samples, the structure was crashed, but the micropores are visible.

Usually, to make porous carbon nanomaterials, templated-assisted synthesis is used. Anuj *et al*. reported that carbon nanosphere derived from oil palm biomass precursors has silica content which helps the material to become porous during pyrolysis. Plants absorb silicon from the soil as $Si(OH)_4$ or $Si(OH)_3O$-, which is then used to produce silica to incorporate into their epidermis and cell walls [7]. Biomass with higher lignin concentrations (grape seeds, cherry stones) produces carbon nanomaterial (CM) with a macroporous structure. In contrast, raw materials with a higher cellulose content (apricot stones, almond shells) produce CMs with a primarily microporous structure [8]. These pores serve different functions [9], with macropores acting as ion reservoirs for micro and mesopores, mesopores providing appropriate channels for the transport of ions during the diffusion process, and micropores acting as molecular sieves that regulate ion diffusion to control the material's capacitance and charge status. Different applications can benefit from these behaviors of porous materials. Physical and chemical treatments are used to decrease particle size and eliminate pollutants from biomass precursors. Powerful acid hydrolysis using HNO_3, H_2SO_4, and HCl is used for chemical treatment, although these strong acids must be neutralized.

Particle size refinement can be done with ball milling and grinding. The mesh is utilized to separate particles of various sizes.

Fig. (1). SEM images of the BCMs prepared from different biomass precursors [6].

The thermochemical process of pyrolysis extracts carbon from biowaste and organic materials. During pyrolysis, the decomposition of organic components and release of the gaseous component from the system will lead to crosslinking of carbons with aromatic ordering [10]. The non-crystalline nature of hemicelluloses and lignins leads to the development of nongraphitic carbons at the pyrolysis temperature [11]. The efficiency of the process is affected by a number of factors, including the composition of the precursor material, the pyrolysis temperature, the amount of time the material spends in the pyrolysis chamber, the particle size, and the physical structure of the material.

2. BIOCARBON FOR SUSTAINABLE ENVIRONMENT

Commercialization of numerous NCs for a wide range of physical and chemical applications is either imminent or has already begun for various carbonaceous materials such as carbon nanodots (CDs), carbon nanofibers (CNFs), and nanodiamonds [12, 13]. BNCs have already made significant advances in a number of applications including batteries [14], supercapacitors [15 - 17], solar cells [18], catalysts [19], electrochemical sensors [20], and structural composites (Figs. **2** & **3**).

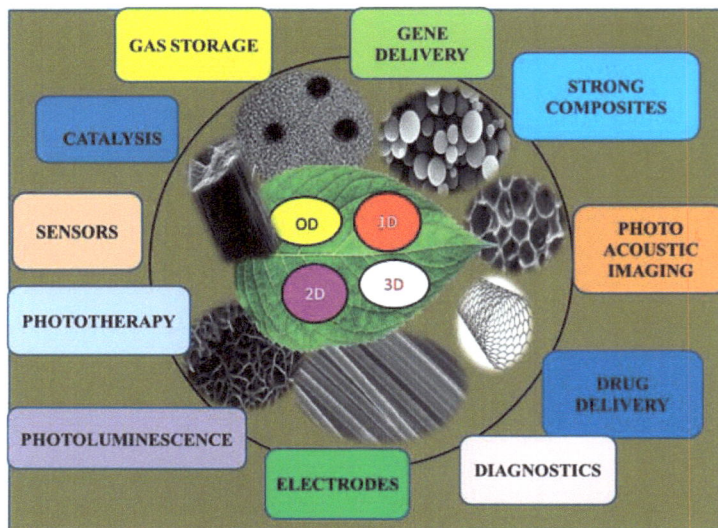

Fig. (2). Application of various biomass precursors.

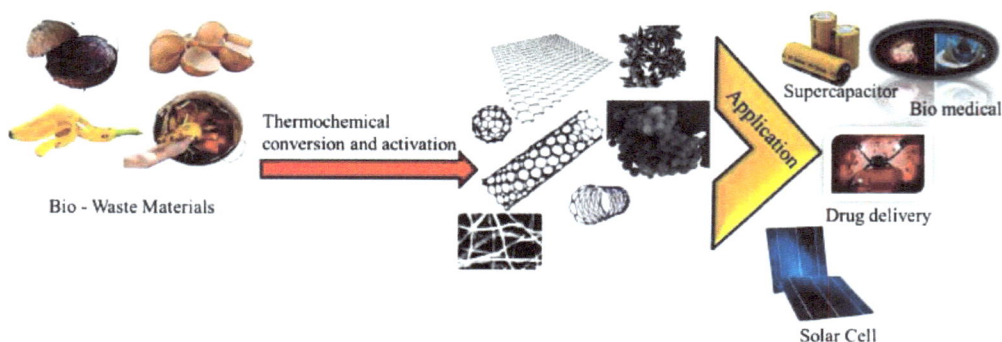

Fig. (3). Application of bionanocarbon nanomaterials.

The widespread application of BNCs in physio-chemical research resulted in their growing popularity in the field of biology. Nano-sized carbon dots, tubes and fibers and graphenes are only a few of the main BNCs that are introduced along with the carbonization of biomass sources [21].

3. BIOCARBON FOR SUPERCAPACITORS

Among the several different energy storage technologies, one of the most advanced and useful devices is supercapacitors (SCs). Biomass-based products have attracted a lot of attention as a potentially sustainable carbon source due to their renewable, abundant, cheap, and environmentally benign characteristics. Other carbon compounds, such as carbon fibers, nanotubes, and carbon generated from biomass, have also been thoroughly investigated and analyzed by prominent research groups [22 - 24] for the development of flexible electrodes. 2D materials like graphene, MXene, a few layers of BN, *etc.*, with polymeric template have been intensively studied for further increment in the efficiency of electrochemical devices [24 - 26]. Recently, carbon forms produced from carbonizing agricultural waste biomass as the active electrode material have drawn a lot of interest because they naturally have a high surface area and hierarchical porosity, two characteristics that are necessary for effective energy storage devices [27 - 29]. Gomaa *et al.* [17] fabricated an electrode for the development of a superior capacitance supercapacitor by making use of mesoporous carbon nanospheres which are produced from biowaste (onion dry peel). The obtained materials show $2962\ m^2g^{-1}$ of surface area and capacitance of 189.4 at 0.1 A/g in 3 M KOH. Vinay *et al.* [19] successfully developed a biomass (garlic peel) derived electrode material for high-performance SCs and the device showed a maximum capacitance of 174 F/g at 0.1 A/g in 4.0 M KOH electrolyte, the specific capacitance of 119.2 F/g at 0.1 A/g, and the capacitance retention of 93% was observed after 10000th cycle. These findings indicate the enormous potential of non-activated carbon nanomaterials generated from biomass for effective, reliable high-performance electrodes in electrochemical energy storage applications.

4. BIOCARBON FOR MEDICAL APPLICATIONS

One of the earliest materials to be used industrially is lignocellulosic biomass, and modern society still uses this resource today. Lignocellulosics have been playing an increasingly significant role in the development of pharmaceuticals since they are also rich sources of the natural ingredients required for traditional medicine. These topics, along with the pharmacological and chemical challenges they present, remain highly prominent and are the subjects of ongoing scrutiny within the scientific community [30].

Suriyaprabha *et al*. [31] studied *in vitro* study on animal cells by making use of non-toxic silica nanoparticles which are derived from bamboo leaves. These nanoparticles showed non-toxic nature up to 125 µg mL^{-1}. The blue-green algae *Spirulina platensis* was used by Kalabegishvili and coworkers to synthesize [32] gold and silver nanoparticles for medical applications. Total gold and silver concentrations found in the biomass revealed that on the first day, the metal ions were quickly absorbed primarily into the cell surface and then slowly moved into bacterial cells. The equilibrium dialysis technique experiments demonstrated the significance of surface processes in the synthesis of metal nanoparticles.

5. THERMAL CONDUCTIVITY STUDY OF BIOCARBON

Woody biomass fuels are in accordance with a general relationship between material density and thermal conductivity, according to data analysis. It has been demonstrated that the compressed solid matter in torrefied wood has a much better heat conductivity than that of untorrefied wood. It has been demonstrated that black and olive residue pellets have higher thermal conductivities than wood, although herbaceous materials often have lower values [33]. Pownraj *et al*. synthesized a low cost environmentally friendly green antioxidant-grafted bio-carbon nanoparticle in order to enhance the tribological properties and thermal conductivity of nanofluids [34]. Hung *et al*. produced biochar from wood waste and incorporated it into a nanofluid using the incipient wetness impregnation method with $Cu(NO_3)_2$, $3H_2O$ and HNO_3 at varying concentrations. The thermal characteristics of nanofluids with about 2% additive content, when the conductivities of [C$_4$mim][Cl] and [C$_4$mim][BF$_4$ thermal] with and without biochar were compared, it was shown that the biochar can improve the thermal conductivities of ILs. This is due to the biochar obtaining its graphitic structure [35].

6. BIOCARBON NANOMATERIALS IN SOLAR CELLS

In order to lower the electrode's price and transporting materials in the development of perovskite solar cells (PSCs), Liguo *et al*., developed an electrode by using four biomass carbon precursors: corn stalk, peanut shell, phragmites australis, and bamboo chopsticks. These prepared electrodes show good photovoltaic performance and PSCs based on bio-carbon. Carbon electrodes (Ces) were more stable than conventional devices. Briscose *et al*. developed a sensitizer for ZnO-nanorod-based solid-state nanostructured solar cells from biowaste-derived CQDs [36].

7. POLYMER NANOCOMPOSITE

Polymers (thermoplastics, thermosets, or elastomers) that have been reinforced with a tiny amount (less than 5% by weight) of nanosized particles (fillers) with a high aspect ratio (L/h > 300) are known as polymer nanocomposites (PNCs) [37]. The main components in a polymer nanocomposite are the polymer matrix, nanoscale fillers reinforced with the polymer, and the materials incorporated to enhance the dispersion [38].

The enthalpic interaction between the filler and polymer influences the stress transfer across the polymer composite. The van der Waals interaction and covalent bond between polymer chains and nanoparticles (filler) define the enthalpic interaction. A large number of fillers have been discovered during the past decades. The chemistry, size, and shape of fillers are significant hallmarks for enhancing polymer nanocomposite properties. The synergy between the polymer and the fillers is inevitable in a polymer nanocomposite study (Fig. **4**).

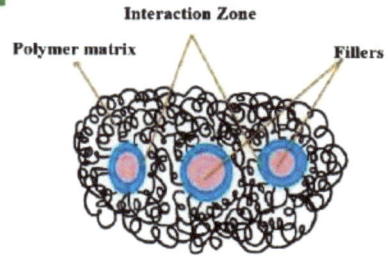

Polymer Matrices for nanocomposites

Thermoset	Thermoplast	Elastomer
• High stiffness & strength • Adhesion Properties	• Better impact resistance • Higher fracture toughness • Moderate stiffness & strength	• Excellent impact resistance • Very high elongation • Low stiffness & strength
• Examples Epoxy Phenolic Polyimide Bismaleimide	• Examples Poly ketones which include Polyether ketone (PEK), Polyether ketone ketone (PEKK)	• Examples Polyurethane

Fig. (4). Polymer matrices for nanocomposite.

8. SYNTHESIS OF POLYMER NANOCOMPOSITE

To create polymer nanocomposites, it is crucial to take into account the following information about the constituents in order to comprehend how the new materials behave: (i) Thermal stability of polymers, (ii) Surface area, (iii) Polymer chemical properties, (iv) Semi-crystallinity of polymers, (v) Polymer mass, (vi) Chemical solubility, (vii) Chemical structure and (viii) Dispersion of nanoparticles. The most popular ways to make these nanocomposites are melt extrusion, dispersion of solution (nanoprecipitation and spray drying), *in situ* polymerization, and solution casting. The interaction and path of the polymer-nanoparticle pair determine the most effective technique. The secret to creating novel materials with compound qualities that synergize is the dispersion and distribution of nanoparticles inside the polymer matrix. The strength of the intermolecular contact between the nanoparticles and the polymer matrix determines how effective this synergism will be. Every process has a unique characteristic. But regardless of the method, the final morphology of all polymer nanocomposites is what matters the most [39, 40]. This depends on interactions between the polymer and the nanoparticles, which will help to evenly distribute and disperse the nanoparticles throughout the polymer matrix.

8.1. *In-situ* Chemical Polymerization

This method is frequently useful for polymers that cannot be made economically or safely using solution methods because the solvents required to dissolve the polymers are excessively toxic. The nanoparticles in the polymer matrix are effectively dispersed and distributed by this approach [41]. It is necessary to highlight a few key components of this approach. The first has to do with the process's expense, which may necessitate certain modifications from the usual polymer synthesis. The best catalyst must be chosen with care as well. The equipment can be used for polymerization with or without nanoparticles [42].

As illustrated in Fig. (**5**), the filler and matrix (polymer) are combined at a high temperature without the use of a solvent, followed by cooling and drying. Nanocomposites with insoluble and thermally unstable matrices (insulating polymers) that cannot be generated by solution/melting procedures are typically created using *in-situ* polymerization.

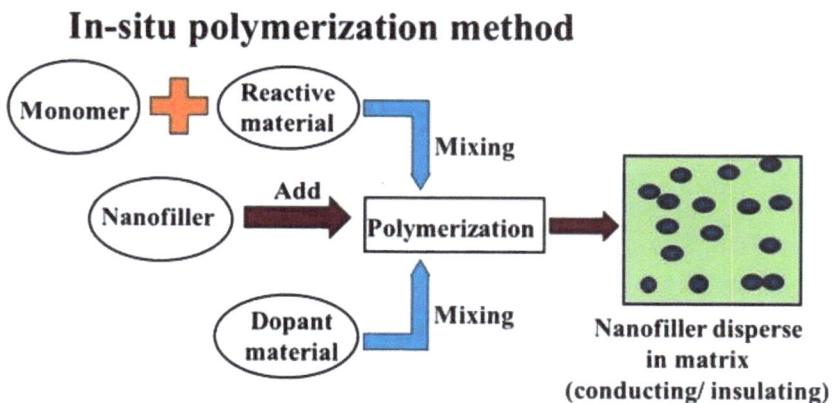

Fig. (5). The process of creating conductive and insulating polymers using *in-situ* polymerization technique [43].

8.2. Solution Method

When the solvent being used is less hazardous, this is a useful technique (chloroform, acetone, alcohol or water). Because the solvent and polymer interact well with the nanoparticles in this approach, variable amounts of them can be disseminated. This is the simplest way to make high-quality nanocomposites. The solvent must be handled carefully because it must be entirely removed thereafter [44]. Combining spray drying with solution dispersion is one intriguing technique. After some homogenization time, the dispersion of nanoparticles in the polymer solution is injected into the spray dryer, which dries the substance into a powder. The final yield of the material is this method's main flaw. However, once all the circumstances have been altered and the end product exhibits good dispersion and dispersal, this process takes place in a single step. This method allows for the injection of drugs for controlled or targeted distribution.

Composite PV membranes are typically created using the solution casting approach. As shown in Fig. (**6**), this procedure involves dissolving the polymer along with all other components to create a homogeneous solution, which is then cast onto a flat glass plate or Petri dish. The dried membrane is then removed from the glass support when the solvent has completely evaporated. The solution casting approach is not only effective for creating single-layered membranes but is also very helpful for creating multi-layered dense and/or porous membranes by using multiple solutions to coat the glass support. In fact, the casting solution can benefit from the addition of highly volatile solvents with an evaporation step before the phase inversion in a non-solvent bath.

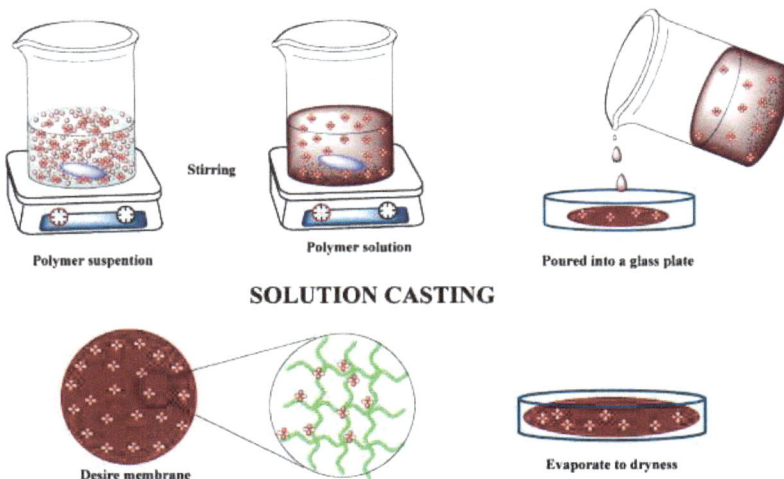

Fig. (6). Solution casting of polymer membrane [45].

8.3. Melt Extrusion

As opposed to the other methods, this one has a significant benefit because no solvent is required. However, it's crucial to consider how many nanoparticles will be disseminated. Due to their propensity to aggregate more readily than in other approaches, this method necessitates careful monitoring of the dispersion of the nanoparticles. The equipment is the same as that used to process polymers without nanoparticles. Therefore, the researcher must be mindful of the temperatures employed to prevent polymer degradation during extrusion, as well as the amount of time required for optimum nanoparticle dispersion. The temperatures required for breakdown and melting are quite similar for most natural polymers and some biopolymers.

By utilizing HME technology, the advancements described here may also enhance how veterinary clinics administer medications. The research also gives a brief summary of the situation of the global market for pharmaceutical HME market from 2016 to 2024. This analysis provides a tabular summary of various medication delivery uses of HME over the last four years (Fig. **7**).

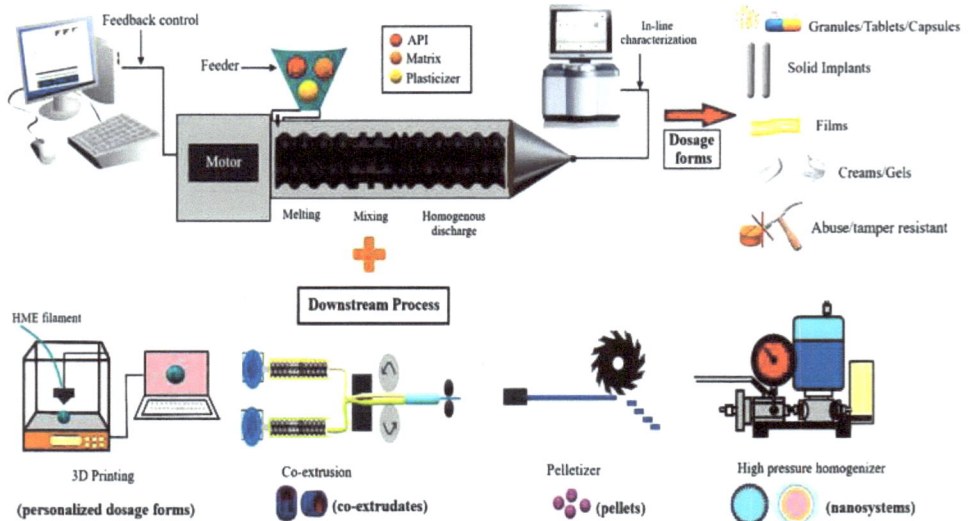

Fig. (7). Basic diagram of an extrusion in pharmaceutical applications [46].

8.4. Sol-gel Method

The sol-gel method is a bottom-up strategy that operates on the polar opposite of all earlier methods. Sol and gel are the first two steps in a sol-gel relationship. The

3D interconnecting network generated between phases is known as gel, and sol is a suspension of solid nanoparticles in a monomer solution in a colloidal fashion. In this method, a monomer solution is used to disperse solid nanoparticles to create a colloidal suspension of solid nanoparticles (sol). By using polymerization and hydrolysis processes to connect the phases, the gel is created. The liquid is completely covered by the 3D network of polymer nanoparticles. The polymer encourages the development of multilayer crystals and acts as a nucleating agent. The polymer seeps between layers as the crystals grow, creating a nanocomposite.

The sol-gel method is used to create inorganic oxide nanoparticles in polysiloxane and epoxy matrices. Several studies conducted over the past two decades have evaluated the effects of the synthesis conditions and the reactants utilized on the inorganic structures generated, the interactions between the polymer chains and the inorganic nanoparticles, and the consequent properties of the nanocomposites. The uses of various polymer-inorganic oxide nanocomposites are also briefly reviewed, along with alternate *in situ* methods (Fig. **8**).

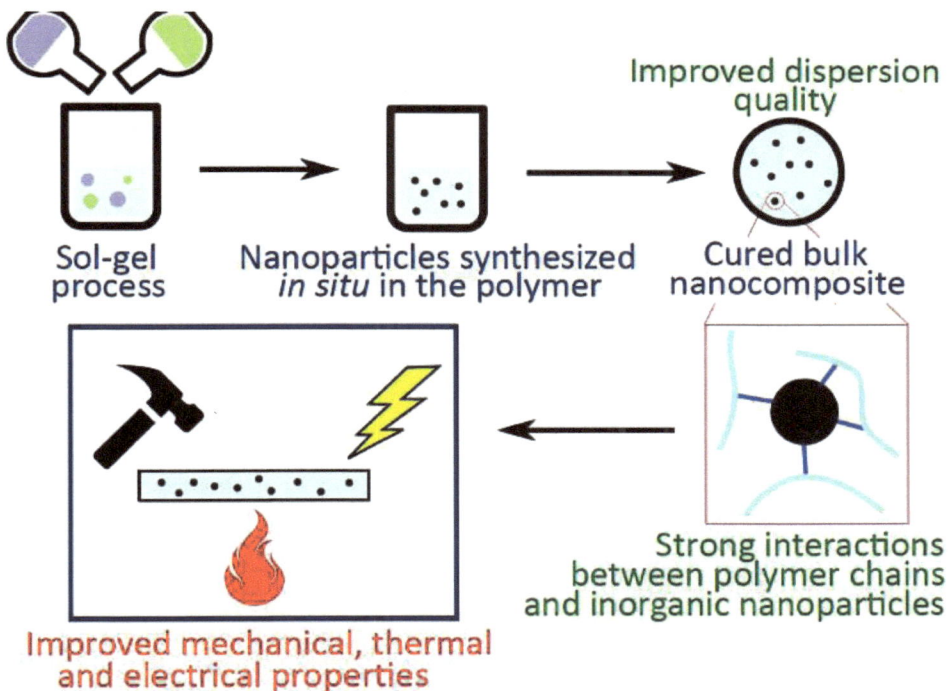

Fig. (8). The sol-gel method to create inorganic oxide nanoparticles in polysiloxane and epoxy matrices [47].

8.5. Melt Blending

The best method for creating clay/polymer nanocomposites with a thermoplastic and elastomeric polymeric matrix is melt blending [48]. Usually, a banbury or an extruder is used to melt the polymer and blend it with the necessary quantity of intercalated clay. An inert gas, such as argon, nitrogen, or neon, is used during melt blending. Alternately, the polymer and intercalant could be mixed dry, heated in a mixer, and then sheared enough to create the required clay polymer nanocomposites. Compared to *in situ* intercalative polymerization or polymer solution intercalation, melt blending has many advantages. Synthesis of inorganic nanomaterials *via* melt blending doesn't use organic solvents, making it environmentally friendly.

By melting 5 wt% of fumed silica in a polycarbonate matrix at a melt temperature of 280 °C and a screw speed of 30 rpm in a LabTech twin screw extruder (Thailand), PC-FS nanocomposite was created. The LabTech pelletizer was used to further pelletize the extruded fibers. Using a hot press technique and a hydraulic press, samples for dynamic mechanical analysis, the tensile test, and the impact test were created in accordance with ASTM standards (ASTM-D4065, ASTM-D638 and ASTM-D256, respectively). Fig. (**9**) depicts the full sample preparation process in detail.

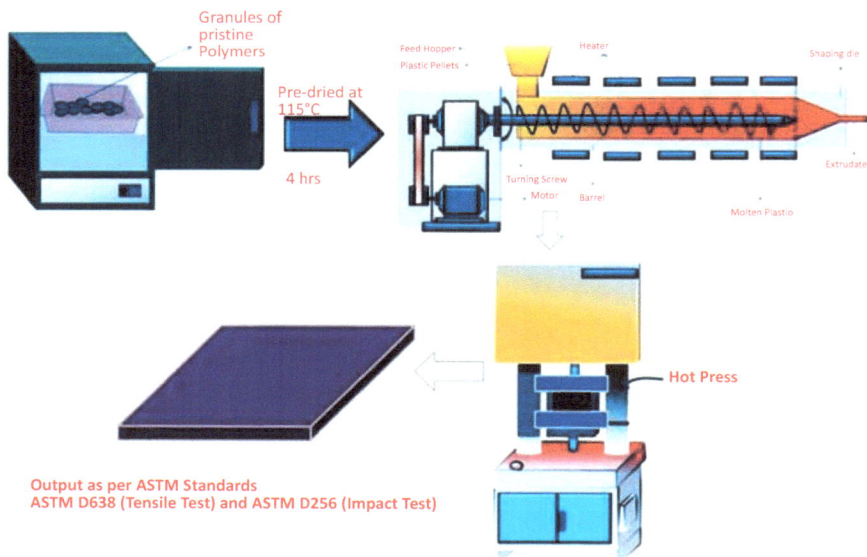

Fig. (9). Fabrications of PC-FS Nanocomposite *via* melt Blending and Hot Press Technique [49].

8.6. Hand Layup Method

The hand layup process, which is illustrated in Fig. (**10a**), involves manually preparing the mold with fabric layers and resin matrix in order to construct the laminate stack. By using hand rollers, the resin in the composite was distributed uniformly in order to prevent air entrapment. The preparation of the mold, covering the gel, laying up the fibers, and post-curing are the procedures in hand layup. Therefore, every step of this operation depends heavily on the operator. The benefits of this approach are its low cost and straightforward operation.

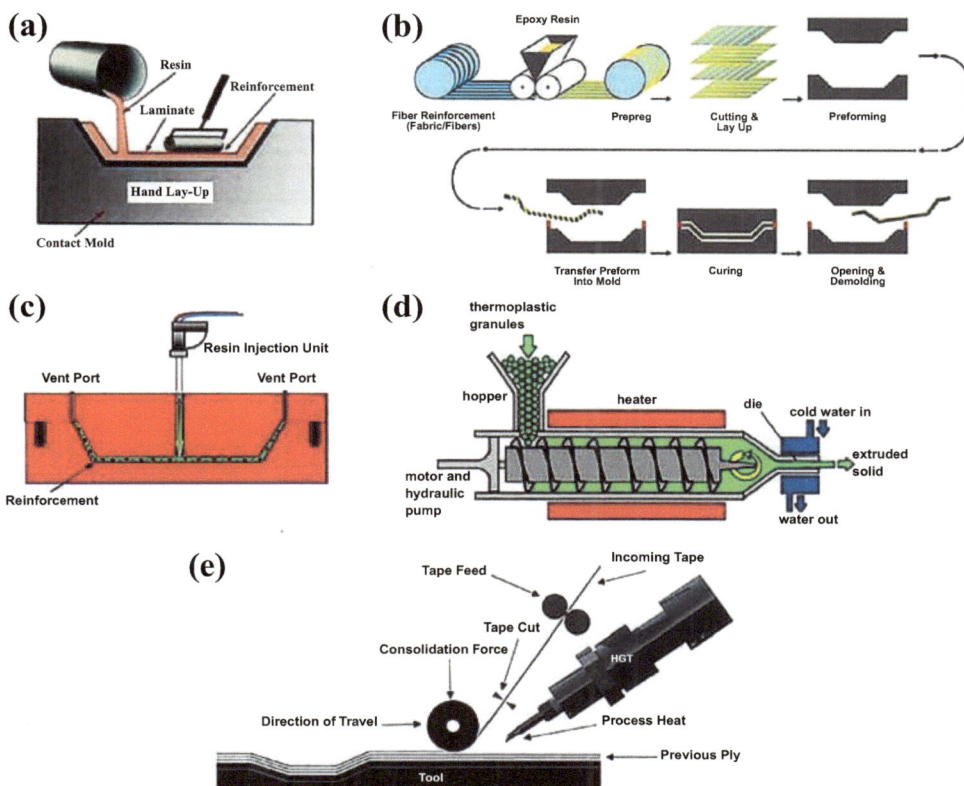

Fig. (10). (a-e) Fabrication Techniques of polymer composite [50].

The selection of an appropriate construction procedure is quite important when creating composite properties. The ideal form that is free of flaws, shape, size, and preferred composite qualities should all be taken into account during the fabrication process, along with production costs, efficiency, and matrix properties [51-53].

9. RECENT ADVANCEMENTS IN THE CARBON BASED POLYMER NANOCOMPOSITE

For applications in lightweight structural polymer composites, carbon-based fillers are emerging as important components with extraordinary appeal. For example, when combined with polymer materials, Carbon fiber has a very strong reinforcing effect and is less in weight than glass fibre (GF) reinforcement. The successful identification of two new carbon allotropes, graphene in 2004 and CNTs in 1991, boosted the development of carbon materials technology even further [54, 55]. Several carbon-based composite materials' mechanical characteristics, including their tensile strength and Young's modulus, are described in Fig. (11). Carbon fibre reinforced polymer composites (CFRP) have mechanical qualities that can exceed 100 GPa of Young's modulus and 100 MPa of tensile strength. For the most part, CNT and graphene-based materials continue to be used often in the creation and development of composites based on carbon since they exhibit the maximum tensile strength and modulus (upper right quadrant of the Ashby chart). However, in actual applications, the manufacturing conditions for these carbon compounds are seldom ever duplicated. The technology itself has lot of short comings such as the use of a lot of chemical solvents, and the hectic purification procedure. Many researchers have sought to take use of CNTs' hollow nano-fiber behavior in composite materials because it results in a significantly reduced weight, high aspect ratios, and remarkable capacities in terms of mechanical, thermal, electric conductivity, and electromagnetic interference. Like other carbon allotropes like graphite, Strong C-C covalent bonds form a ring structure in the sp^2 configuration in CNTs. Although the CNT particles have special characteristics, they are not susceptible to chemical processes or matrix-interface adhesion. Because of this, CNTs require further processing, such as surface modifications, to boost their affinity for plastic surfaces [56]. Despite extensive research on CNT, graphene, and related derivatives as useful polymer fillers, there are yet few practical uses for these composite materials. Owing to their complex, harsh chemical preparation procedure, which results in extremely expensive end-goods. It is still challenging to streamline and lower the cost of producing these high-performance carbonaceous materials [53].

The search for a replacement and less expensive filler sources, biomass material, has become imperative. Because of their availability, thermal stability, and low energy demand for manufacture, biomass-derived carbon nanomaterials have been regarded as an excellent material for polymers' mechanical strength and thermal enhancement [52].

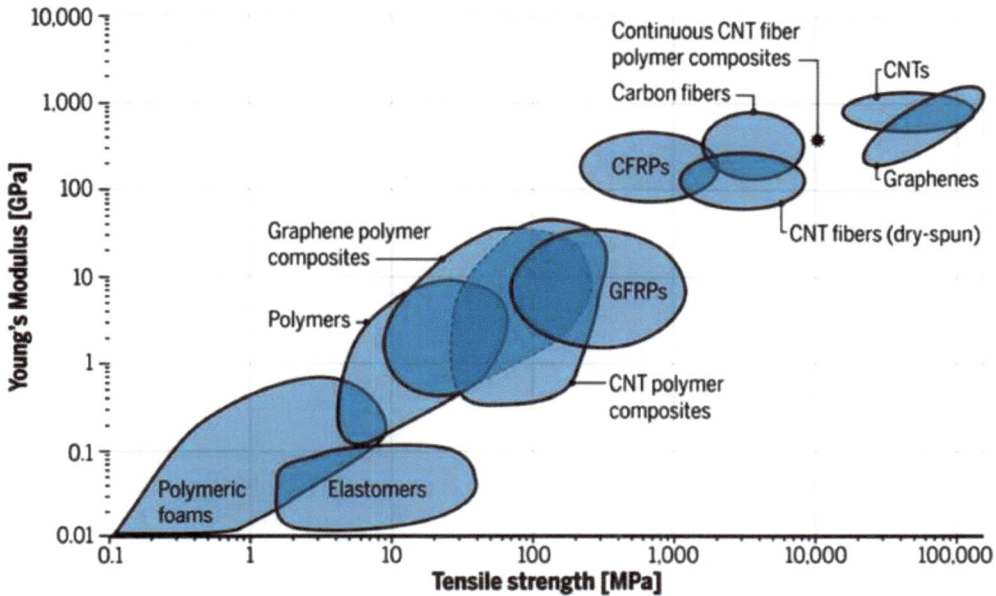

Fig. (11). Tensile properties for different modern carbon-based materials reinforced polymer composites [57].

The utilization of hybrid systems that permit the input of large amounts of BC meeting certain industrial criteria is described in US patent 9809702B2 [58]. In particular, the combination of natural fiber, GF, or CF with BC is covered in this document. By its very nature, BC provides the black color to the textiles without the use of Carbon black. Composites with a tiny amount of CF and a mixture of BC are said to function exceptionally well. The necessary performance characteristics for lightweight automotive components were given by hybrid BC fillings with low concentrations of GF and CF. Carbon-reinforced polymers have various applications, including biomedical, sensors and actuators, defense and aerospace, antimicrobial, environmental cleanup, food packaging, and water remediation [59]. Here in the book chapter, we focus on the enhancement of the properties of various types of polymers embedded with bio-nanocarbon.

10. SYNERGISTIC EFFECT OF BIO-NANOCARBON AND POLYMERS

10.1. Biocarbon Reinforced Thermoset Polymers

Particularly tempting as reinforcing agents in thermoset resins for high-performance applications are carbon-based fillers. Thermosets are a particular family of polymers that cure with the addition of a curing agent or by heating to create well-defined, irreversible chemical networks that have a tendency to develop in three dimensions. The usage of thermosets is widespread in the building and construction sector, as well as in electrical equipment, adhesives, and transportation. Moreover, thermosets are widely employed in sophisticated applications, particularly in the aerospace and military sectors. There are reports on the reinforcement of BC into the thermoset matrix like epoxy resin, polyurethane, and polyester resin that exhibited the enhancement of the properties of thermoset polymers due to the synergistic effect of polymer and embedded biocarbon.

10.2. Epoxy Resin

Epoxy resin is a thermoset polymer matrix that has varieties of applications in the aerospace and automobile industries. The exceptional qualities of epoxy resin include excellent adherence to various materials, inertness to harsh conditions, minimal shrinking, and good electric and thermal qualities. Epoxy resin is a catch-all term for substances containing two or more oxirane rings (epoxy groups). A suitable curing agent cures the molecules in three dimensions. They are created by the combination of bisphenol A and epichlorohydrin, which is the most important precursor in the synthesis of epoxy resin [60]. Because of the oxirane ring that gives the molecule its flexibility, epoxy resins behave as surfactants and are chemically well-suited to most substrates. Epoxy resin is also easily cured without the formation of by-products or volatiles, making it excellent for composite applications. In structural applications, epoxy resin is particularly brittle and notch-sensitive, prompting further research to enhance toughness.

The mechanical characteristics of thermosetting materials are improved when fillers like coconut shell powders are added. Coconut shell powder-based polymer composite showed enhancement in flexural strength, impact strength, and hardness [23]. Agunsoye *et al*. studied the properties of carbonized coconut shell nanoparticles reinforced with epoxy resin. They reported that tensile strength at break and flexural strength of epoxy resin is improved with the amount of carbon nanofiller increased by 5-25 wt%. The material showed higher values at 25 wt% of carbon particles, the higher values are due to the effective cross-linking between the matrix and the polymer [61]. The mechanical performance of epoxy-based biocomposites is significantly influenced by the shape of the BC produced from various biomass feedstocks. The ductility of BC with a flat surface increases, whereas surfaces with significant porosity and irregularities become more brittle [62].

The epoxy composite prepared from oil palm leaves derived carbon nanospheres (CNS) was another material that showed enhancement of properties. The composites manufactured with 0.25 wt% CNS-reinforced composites show notable improvements in ultimate tensile strength and Young's modulus when compared to normal epoxy (strain). The SEM analysis of the structure confirms the effective dispersion of nanoparticles due to strong interfacial bonds in the polymer matrix. The toughness of 0.1 wt% and 0.25 wt% CNS loaded samples improved by approximately 15% and 150%, respectively, compared to clean epoxy. For 0.1 and 0.25 wt% CNS-epoxy nanocomposites, Young's modulus rises by 3.2% and 11%, respectively. The dampening capability of 0.25 wt% CNS-epoxy is reduced by 14% due to particle agglomeration (Fig. **12**) (Table **1**).

Fig. (12). The fabrication of biocarbon/epoxy composites [63].

Table 1. List of examples of bio nanocarbon embedded epoxy resin.

S. No	Bio Carbon source	Technique	Enhancement of Properties	Refs.
1.	Coconut shell powder (CNP of about 50 nm)	Hand lay up	338.75 MPa of tensile stress and 156 MPa of flexural strength were measured in a 25% composite.	[61]
2	Arecanut shell and coconut shell powder	Hand lay up	Density decreased, increase in void content and water absorption capacity observed	[64]
3	A cluster of oil palm fruit nanofibers.	Hand lay up	The highest values of the storage modulus, loss modulus, and damping factor were obtained with the addition of 3 wt% nanofillers.	[65]

(Table 1) cont.....

S. No	Bio Carbon source	Technique	Enhancement of Properties	Refs.
4	Groundnut Shell Powder	Hand lay up	Tensile strength was about 23 MPa at 15% filler loading; flexural strength was approximately 53 MPa at 15% filler loading composite; hardness rose as filler loading wt% increased.	[66]
5	Apple wood, coconut powder, and apple shell	Hand lay-up	Strong tensile and compressive performances were achieved by composites comprising 15% wood apple shell powder and 25% coconut shell powder, Comparatively, composites comprising 27% wood apple shell and 3% coconut shell had enhanced flexural characteristics. Only 30% of wood apple shell-loading produced satisfactory results in impact strength testing.	[29]

10.3. Biocarbon-filled Polyurethane (PU) and Polyester Resin

CB-reinforced PU is ubiquitous in automotive industries. Low molecular weight glycols find their primary applications as chain extenders, diisocyanates, and polyols, constituting the essential trio of raw materials utilized in the production of polyurethane (PU). The typical source of these components is petrochemicals, although researchers have recently discovered biobased substitutes. The environmental benefits of using bio-resources include lower energy consumption during manufacturing, fewer greenhouse gas emissions, lower CO_2 emissions, and enhanced biodegradability, particularly when bio-based polyester polyols are employed [67]. Biocarbon derived from lignin was used to synthesize the PU foam for dye absorption applications. It exhibited high adsorption towards cationic dye. CB is often used to strengthen polyurethanes. Due to BC's polar nature, there are tremendous prospects for its application as filler in polyurethanes. The utilization of environmentally friendly biobased polyurethanes, biofillers, and BC is a fascinating subject that needs additional study [68].

10.4. Polylactic Acid (PLA)

As a bio-based and biodegradable polymer for the thermoplastic sector, poly (lactic acid) (PLA) remains a viable choice, although paths are still being studied for more extensive uses. In contrast to the traditional areas of attention, which include single-use products and short-term packaging, there is potential for its usage in composite formulations for more specialized applications such as automotive components, electronics, durables, and long-term packaging [70, 71]. PLA's drawbacks include low thermo-mechanical stability, weak toughness, and inadequate gas barrier properties [72], which restrict its end-use applications when biodegradability or bio-content are desired. One method to mitigate these issues is by employing blending techniques or copolymerization [73]. However, filler variants have been incorporated into PLA to act as an industrially and economically reasonable solution by reducing the cost of the composite while strengthening targeted properties such as dimensional stability, strength or stiffness, permeability, and fire retardance features [74]. Here we are mainly concerned about the effect of incorporating biocarbon as fillers into the polylactic acid matrix. The biocomposite obtained from spent coffee grounds and polylactic acid through melt mixing and solvent casting showed improvement in the tensile properties of the composite. The composite synthesized from solvent casting exhibited a crystalline nature and the composite result from melt mixing showed an amorphous structure. The findings also suggest that reducing the size of biocarbon particles will improve the interactions between the filler and the polymer, resulting in more effective dispersion within the composite [75]. For a polymer composite application, a filler or reinforcement's surface hydrophobicity is crucial. It determines how well the filler and matrix work together, which affects mechanical performance. PLA with wheat stem BC pyrolyzed at 800 showed good wetting and interfacial bonding. For a polymer composite application, a filler or reinforcement's surface hydrophobicity is crucial. It determines how well the filler and matrix work together, which affects mechanical performance. PLA with wheat stem BC pyrolyzed at 800^{0}C showed good wetting and interfacial bonding [6]. Yeng *et al.* reported that the reinforcement of banana fiber improved the PLA's thermal stability and mechanical properties. The tensile and flexural strength of the composite is 2 and 1.666 times higher than the pristine polymer. In addition, the HDT of the composite improved by 122% at 40 wt% banana fiber content. The bio-nano carbon powder of particle size 21.7 nm is used as a filler material in PLA matrix and has been applied as biodegradable antistatic packaging material. The PLA-based nanocomposite was prepared by solvent casting method containing 0.5-10 wt% BC material. The conductivity of the material was observed to be increasing as the biocarbon particle amount. Contact angle study of the material exhibited high surface wettability compared with neat PLA, indicating that the material can relieve and enhance the decay of static electricity and promote electronic device performance [76]. Polymer affinitive photoluminescent quantum dots are prepared from the spent coffee ground through a microwave-enabled synthesis method. With only 0.5 weight percent QDs, PLA nanocomposites inherit the special optical characteristics of QDs, resulting in intense luminescence, high visible transparency, and a nearly 100% UV-blocking ratio in the entire UV region. Strong PLA chain anchoring at the QDs' nanoscale surfaces made it easier for PLA to crystallize, which resulted in appreciable gains in its Tensile and thermomechanical characteristics. For instance, the tensile strength and storage modulus at 100 °C both rose by about 2500% and 69%, respectively, using 1 wt% QDs, in comparison to those of pure PLA (4 MPa and 57.3 MPa). The energy-dissipating and flexibility-imparting mechanisms enabled by QDs during tensile deformation, PLA nanocomposites displayed an uncommon combination of elasticity and extensibility as a result of phenomena such the creation of multiple shear bands, crazing, and nanofibrils. This creates the opportunity for biowaste-derived nanodots with strong polymer affinity to apply exquisitely differentiated light control and extreme nano reinforcing in an eco-friendly way [77] (Figs. **13** & **14**).

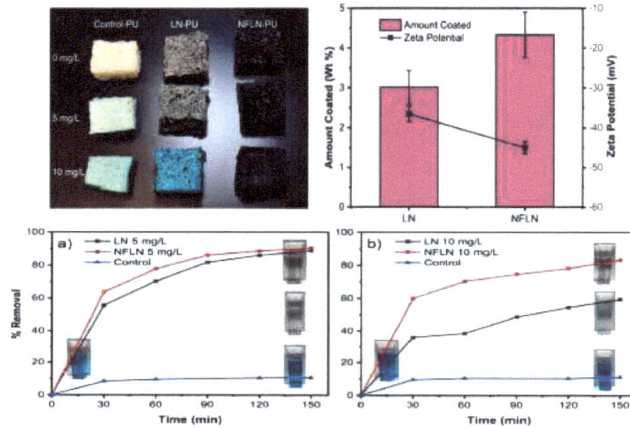

Fig. (13). PU/BC system before and after dye adsorption and embedded biocarbon zeta potential data. a) And b) showing dye adsorption data at different concentrations of methylene blue dye [69].

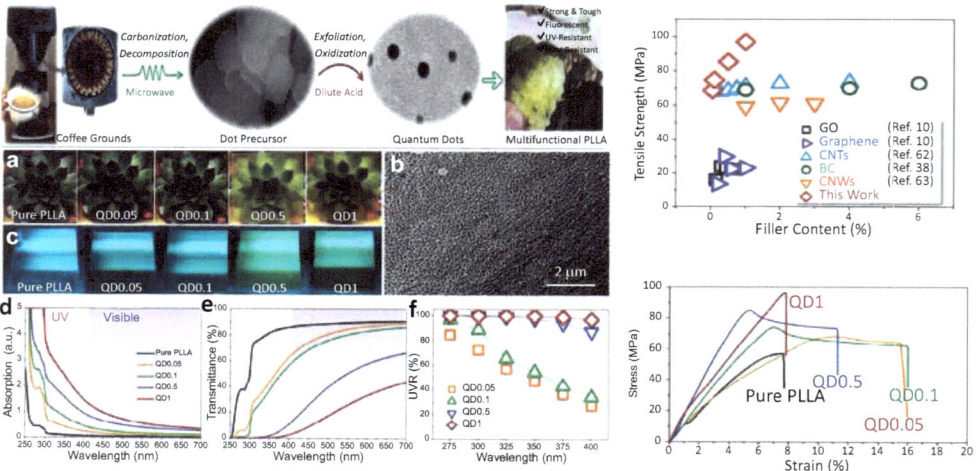

Fig. (14). Schematic illustration of the synthesis of coffee grounds derived from QDs and multifunctional biocomposite using PLA as a matrix. The prepared QD/PLA nanocomposite exhibited good optical properties and enhanced mechanical properties [77].

Mei. *et al.* made bamboo charcoal reinforced with PLA. The prepared bamboo charcoal particle had a uniform surface area and similar aspect ratio, which attributed to the improvement of tensile strength and flexural strength by 43% and 99%, respectively. Furthermore, the ductility index of the composite material is improved by 52% compared with pristine PLA. Bamboo charcoal-reinforced material is resistant to degradation in the presence of UV and compost degradation [78]. Nanocomposite fibers are developed by the electrospinning PLA and groundnut shell powder (GSP) solution. The qualities like hardness, stiffness, ultimate tensile strength, energy, and stress at break of PLA reinforced with 5 wt% untreated GSP showed the best mix, while ductility was the only medium. In the SEM image of the electrospun fibers with higher filler content, small beads are observed due to the agglomeration of filler particles [79] (Fig. **15**) (Table **2**).

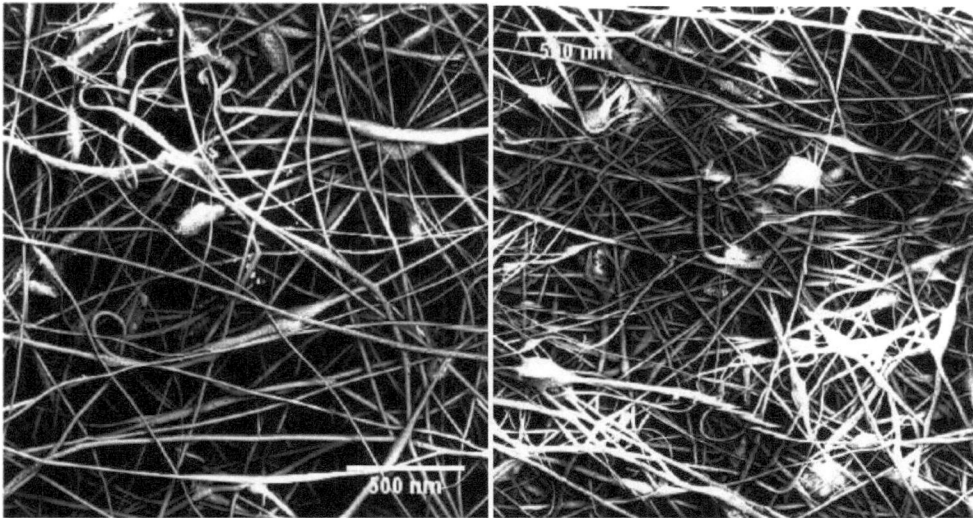

Fig. (15). The morphologies of the electrospun fibers formed from PLA-GSP solution [79].

Table 2. List of techniques used for bio mass precursors to obtain nanocarbon.

S. No	Biomass Precursor	Technique	Enhancement of Properties	Refs.
1	Sugarcane bagasse	Co-rotating twin-screw extruder	The filler loading of 2 wt% showed a maximum tensile strength of about 98% and flexural strength of about 93.91%.	[80]
2	Groundnut shell powder	Solvent mixing	The ultimate tensile strength of nanofibers comprising 5% unprocessed GS particles was 0.85 MPa.	[79]

11. CONDUCTING POLYMERS

Polymers were thought to be an electrical insulator until the invention of conducting polymers. However, these polymers have unique optical and electric properties like inorganic semiconducting materials [81]. The electrical and optical properties of conducting polymers are due to the highly polarized, conjugated, delocalized electron-dense bonds present in the polymeric chain. Due to their wide range of uses in things like energy storage, electromagnetic absorption, enhanced transistors, optical limiting devices, biosensors, sensors, and advanced transistors, conducting polymer-based materials have drawn a lot of attention [82]. Polyacetylene (PA), polyaniline (PANI), polypyrrole (PPy), polythiophene (PTH), poly(para-phenylene) (PPP), poly(phenylenevinylene) (PPV), Polypropylene(PP) and polyfuran are examples of common conducting polymers (PF). Functional groups in conducting polymers have pseudocapacitance properties, and as a result, the substance itself exhibits conductivity. In addition to the electrical double layer (EDL), they are also capable of storing charge *via* a quick faradic charge transfer. Due to these factors, the specific capacitances of conducting polymers are higher than those of EDLCs made of carbon materials. However, conducting polymers have enormous properties, and the durability of the material is poor. Moreover, the cycling performance of this material is low when used on supercapacitors. It is the biggest challenge faced by this material. Reinforcement of carbon material into the polymer matrix is an efficient way to solve this problem. It will solve not only the supercapacitor efficiency but also the durability issues. This is mainly due to the synergistic effect between the filler material and polymer matrix. Utilizing carbon fillers derived from biomass offers cost-effective and eco-friendly alternatives for enhancing the performance of polymer-based nanocomposites. Using an easy-to-use yet innovative self-assembly process and an ice template, the N-self-doped carbon structure was assembled from chitosan. The approach featured a hierarchical porous structure and did not require any kind of physical or chemical activation. The porous nature of the biocarbon helped in the infiltration and efficient coating of PANI. It also helped in the rapid and efficient adsorption and desorption of electrolytes. Due to the composite electrode's short diffusion channel, uniformly coated PANI, and ease of access to electrolytes, it possessed a very high super capacitance of 373 F g^{-1} (1.0 A g^{-1}) and outstanding rate capability (275 F g^{-1}, 10 A g^{-1}). The symmetric supercapacitor demonstrated a high super capacitance of up to 285 F g^{-1} (0.5 A g^{-1}) and an extremely high energy density of 22.2 Wh kg^{-1}. In addition to that, composite also has good cycling stability [83]. It was reported that a binder-free super capacitor fabricated from PPy incorporated carbon aerogel. Carbon aerogel was prepared from the biomass cattail. The fabrication of carbon aerogel/PPy composite was through the *in-situ* oxidation-polymerization method. The areal capacitance exhibited by the carbon aerogel/ PPy composite was 419 mFcm^{-2} [84]. Das *et al.* reported that pine wood-derived biochar-reinforced polypropylene biocomposite exhibited good tensile strength and flexural modulus. It was observed that the high surface area of biochar (335 m^{2}g^{-1}) helps in the mechanical interlocking with polypropylene and improves mechanical properties. A lower peak heat release rate (in comparison to neat PP) is seen in the biocomposites when neat PP is mixed with biochar (up to 35 wt%). By preventing heat and mass transmission between the surface and the underlying PP, the compact char layer that forms after burning effectively improves the insulating qualities of the biocomposites [85] (Fig. **16**).

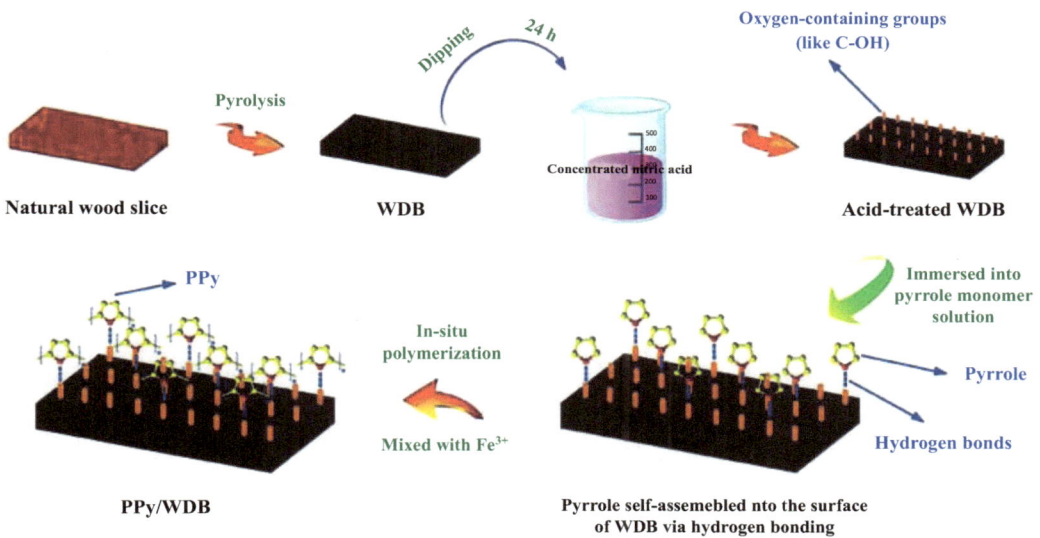

Fig. (16). Process flow diagram for fabrication of composite electrodes made from polypyrrole and biocarbon [86].

Biocarbon fibers made from Douglas fir pulp were combined with a polyamide 12 matrix to create composites. Hot compression mounting was used to create the composites. The composites exhibited log electrical conductivity values of 6.67 S/cm at a filler loading of 7.5 wt% and 0.31 S/cm at a filler loading rate of 35 wt%. In comparison to the composites filled with biochar fiber at loading rates of 25 and 35 wt%, The electrical conductivity values for the samples containing 20 and 40 weight percent carbon fibers are much lower, indicating that the biochar filler is effective as a conductive filler for producing electrically conductive, workable composites [87]. Simple carbonization and *in situ* oxidative polymerization were effectively used to create composite hierarchical structures made of biomass carbon from loofah that were embellished with aligned polyanilines (PANI).In the frequency range of 2-18 GHz, the alignment of PANI and BC was tested for its microwave absorption capability. The outcomes showed that aligned PANI/BC having 30 wt% filler loading had more distinct dielectric response characteristics and improved microwave absorption ability [88].

12.　APPLICATIONS　OF　BIOCARBON-BASED　POLYMER NANOCOMPOSITE

Due to their strength, lightweight nature, pliable processing behavior, and cost-effectiveness, polymer matrix-based composites play a crucial role in numerous industrial and technical processes. The industry's demand for polymer composites to replace metal is driving research into the creation of polymer-based composites with improved characteristics. For applications involving sustainable materials, bio nanocarbon was examined as a material to improve the properties of fiber-reinforced nano biocomposite. Currently, A thorough investigation has been conducted using nano biocarbon as a filler and reinforcing material. However, the composite's poor mechanical, physical, and thermal properties are a result of a poor fiber-matrix interface. As a result, Samsul *et al.* synthesized a biocarbon nanoparticle and used it to improve the characteristics of composite materials and develop functional materials. The oil palm shell from which the bio nanocarbon was made was used to improve the vinyl nonwoven kenaf fibre composite [89]. Selvaraj *et al.* attempted to develop carbon-based dairy farm waste-based hetero-structured biobased benzoxazine benzoxazine-epoxy matrix composites. Studies on the antimicrobial and anticorrosion properties of the prepared cow dung carbon (f-CDC) reinforced hetero-structured biobased benzoxazine-epoxy composites and hetero-structured benzoxazine-epoxy (HSBBz-EP) composites were carried out to confirm the distinctive qualities of the composites suitable for advanced applications.

The composite sample has a high glass transition temperature (199.90), better hydrophobic behavior (contact angle of water is 113.60), corrosion resistance improved by (99.4%) in 3.5 weight % sodium chloride solution, improved antibacterial activity (10mm), and higher thermal stability (293.8°C) with 5% weight loss [90]. Biopolymer-based nanocomposites are a novel class of material with significantly improved mechanical, thermal, and barrier properties. They have been proposed as novel and alternative packaging materials [91]. Chibu *et al.* synthesized a biodegradable plastic using a mixture of BIOPLAST GF 106/02/PLA 75/25 blend (BPB) and carbon nanoparticles (CCSP10) produced from discarded coconut shells. The BPB polymer composite filaments with carbon nanoparticle infusions of 0.2, 0.6, and 1 weight percent have enhanced in mechanical strength. Tensile strength increased by 50% compared to pure BP filaments with the addition of carbon nanoparticles up to 0.6 wt% but later decreased because of particle aggregation [92]. Nanomaterials or nanocomposites can be used as antimicrobial carriers, growth inhibitors, antimicrobial packaging films, antimicrobial agents, or antimicrobial materials. Specifically for cheese, meat, bread, fish, poultry, vegetables, and fruits, antimicrobial nanocomposite films are used as food packaging materials [93] (Figs. **17** & **18**).

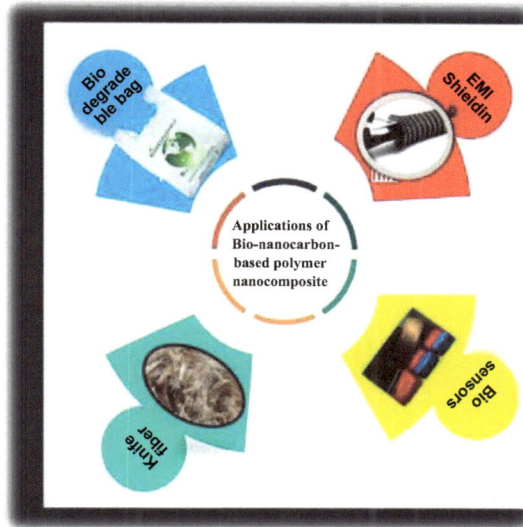

Fig. (17). Applications of Bio-nanocarbon-based polymer nanocomposite.

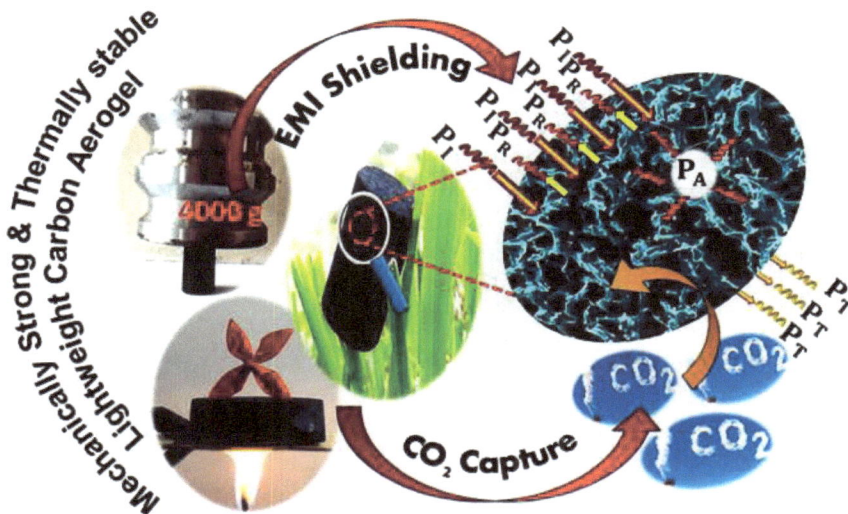

Fig. (18). Biomass-derived carbon aerogel reinforced in PVA for EMI shielding and CO_2 absorption [94].

In this digital age, high-performance electromagnetic interference (EMI) shields are crucial for shielding people and high-tech electronics from dangerous EM radiation emissions. Devi *et al.* developed a doped new cobalt-onion peel-based polyvinyl alcohol (PVA) composite for testing the effectiveness of the material's EMI shielding in high-frequency bands like the X and Ku region. For the purpose of shielding electronic devices from the EMI effect, a flexible electromagnetic shielding material was developed in this study. The cobalt/chopped carbon fiber (CCF) was combined with the biochar particles made from red onion peel to create a composite structure. The results showed that adding charcoal and CCF increased the relative permittivity by up to 9.6. For the PVA composite with 2 vol% cobalt added, the hysteresis analysis also revealed a broad "S" curve [95]. A unique multimedia laminated system was developed by Daniel *et al.* using layered double hydroxide/multiwalled carbon nanotube hybrid bio-nano-composites [NR/St-LDH/MWCNT]. -Ag-[NR/St-LDH/MWCNT] Layers of silver nanosheets. The synthesized nanocomposite displays an incredibly high electromagnetic interference shielding efficiency (EMI-SE) of 28 dB in the absence of a silver sheet and an increased EMI-SE of 84 dB in the presence of a silver nanosheet. A sliver for the tripling process was found in the silver-sandwiched system's absorption efficacy, which was found to be 99.999%. Future EMI shielding applications can use the proposed hybrid material as a highly intelligent, lightweight shielding material, according to thorough physio-chemical and EMI shielding effectiveness evaluations [96].

CONCLUSION

The nano-carbon reinforced polymer matrix composites with bio-inspired design offer exceptional mechanical, electrical, and thermal properties. They often have high strength and toughness at the same time, which fixes the issue of general composites' growing strength and declining toughness and has a promising future for use. It has been developed to use bio-carbon nanoparticles as functional parts and to enhance the characteristics of composite materials. Recent research in the field of bio-inspired nano-carbon composites is presented in this article in order to highlight the technical and scientific challenges that still need to be overcome in order to produce artificial composites with bio-inspired structures. Polymer composites can acquire unexpected properties and novel functions based on synergistic effects when different combinations with various structures, geometries, and properties are added to the composites. Currently, the disciplines of electronics, manufacturing, transportation, and aerospace have all seen extensive research and application of composite materials.

ACKNOWLEDGEMENT

One of the authors Vandana M would like to thank the Centre for Research, CHRIST (Deemed to be University) for providing the post-doctoral fellowship.

REFERENCES

[1]   Roondhe, B.; Sharma, V.; Saxena, S. Theoretical and Computational Investigations of Carbon Nanostructures. In: *Carbon Nanomaterial Electronics*; Devices and Applications, **2021**; pp. 139-154.

[2]   Mohamad Nor, N.; Lau, L.C.; Lee, K.T.; Mohamed, A.R. Synthesis of activated carbon from lignocellulosic biomass and its applications in air pollution control : A review. *J. Environ. Chem. Eng.,* **2013**, *1*(4), 658-666.
[http://dx.doi.org/10.1016/j.jece.2013.09.017]

[3]   Zhang, B.; Jiang, Y.; Balasubramanian, R. Synthesis, formation mechanisms and applications of biomass-derived carbonaceous materials: A critical review. *J. Mater. Chem. A Mater. Energy Sustain.,* **2021**, *9*(44), 24759-24802.
[http://dx.doi.org/10.1039/D1TA06874A]

[4]   Liu, W.J.; Jiang, H.; Yu, H-Q. Thermochemical conversion of lignin to functional materials: A review and future directions. *Green Chem.,* **2015**, *17*(11), 4888-4907.
[http://dx.doi.org/10.1039/C5GC01054C]

[5]   Wang, Y.; Zhang, M.; Shen, X.; Wang, H.; Wang, H.; Xia, K.; Yin, Z.; Zhang, Y. Biomass-derived carbon materials: Controllable preparation and versatile applications. *Small,* **2021**, *17*(40), 2008079.
[http://dx.doi.org/10.1002/smll.202008079]

[6]   Ertane, E.G.; Dorner-Reisel, A.; Baran, O.; Welzel, T.; Matner, V.; Svoboda, S. Processing and wear behaviour of 3D printed PLA reinforced with biogenic carbon. *Adv. Tribol.,* **2018**, *2018*, 1-11.
[http://dx.doi.org/10.1155/2018/1763182]

[7]   Kumar, A.; Hegde, G.; Manaf, S.A.B.A.; Ngaini, Z.; Sharma, K.V. Catalyst free silica templated porous carbon nanoparticles from bio-waste materials. *Chem. Commun.,* **2014**, *50*(84), 12702-12705.
[http://dx.doi.org/10.1039/C4CC04378B] [PMID: 25100105]

[8]   Bogardi, I.; Kuzelka, R.D.; Ennenga, W.G. *Nitrate Contamination: Exposure, Consequence, and Control*; Springer: Berlin, Heidelberg, **2013**, p. (30)520.

[9]   Li, W.; Zeng, M.; Shao, Y.; Fan, X. Advanced materials for supercapacitors. *Frontiers Media SA,* **2020**.
[http://dx.doi.org/10.3389/978-2-88963-893-2]

[10]   Yang, K.; Gao, Q.; Tan, Y.; Tian, W.; Qian, W.; Zhu, L.; Yang, C. Biomass-derived porous carbon with micropores and small mesopores for high-performance lithium-sulfur batteries. *Chemistry,* **2016**, *22*(10), 3239-3244.
[http://dx.doi.org/10.1002/chem.201504672] [PMID: 26807663]

[11]   Happi Emaga, T.; Robert, C.; Ronkart, S.N.; Wathelet, B.; Paquot, M. Dietary fibre components and pectin chemical features of peels during ripening in banana and plantain varieties. *Bioresour. Technol.,* **2008**, *99*(10), 4346-4354.
[http://dx.doi.org/10.1016/j.biortech.2007.08.030] [PMID: 17931857]

[12]   Yeamsuksawat, T.; Morishita, Y.; Shirahama, J.; Huang, Y.; Kasuga, T.; Nogi, M.; Koga, H. Semicarbonized subwavelength-nanopore-structured nanocellulose paper for applications in solar thermal heating. *Chem. Mater.,* **2022**, *34*(16), 7379-7388.
[http://dx.doi.org/10.1021/acs.chemmater.2c01466]

[13]   Power, A.C.; Gorey, B.; Chandra, S.; Chapman, J. Carbon nanomaterials and their application to electrochemical sensors: A review. *Nanotechnol. Rev.,* **2018**, *7*(1), 19-41.
[http://dx.doi.org/10.1515/ntrev-2017-0160]

[14]　Yu, K.; Wang, J.; Wang, X.; Liang, J.; Liang, C. Sustainable application of biomass by-products: Corn straw-derived porous carbon nanospheres using as anode materials for lithium ion batteries. *Mater. Chem. Phys.,* **2020**, *243*, 122644.
[http://dx.doi.org/10.1016/j.matchemphys.2020.122644]

[15]　Zhou, H.; Zhan, Y.; Guo, F.; Du, S.; Tian, B.; Dong, Y.; Qian, L. Synthesis of biomass-derived carbon aerogel/MnO composite as electrode material for high-performance supercapacitors. *Electrochim. Acta,* **2021**, *390*, 138817.
[http://dx.doi.org/10.1016/j.electacta.2021.138817]

[16]　Bhat, V.S.; Krishnan, S.G.; Jayeoye, T.J.; Rujiralai, T.; Sirimahachai, U.; Viswanatha, R.; Khalid, M.; Hegde, G. Self-activated 'green' carbon nanoparticles for symmetric solid-state supercapacitors. *J. Mater. Sci.,* **2021**, *56*(23), 13271-13290.
[http://dx.doi.org/10.1007/s10853-021-06154-z]

[17]　Ali, G.A.M.; Supriya, S.; Chong, K.F.; Shaaban, E.R.; Algarni, H.; Maiyalagan, T.; Hegde, G. Superior supercapacitance behavior of oxygen self-doped carbon nanospheres: A conversion of Allium cepa peel to energy storage system. *Biomass Convers. Biorefin.,* **2021**, *11*(4), 1311-1323.
[http://dx.doi.org/10.1007/s13399-019-00520-3]

[18]　Li, J.; Yun, S.; Han, F.; Si, Y.; Arshad, A.; Zhang, Y.; Chidambaram, B.; Zafar, N.; Qiao, X. Biomass-derived carbon boosted catalytic properties of tungsten-based nanohybrids for accelerating the triiodide reduction in dye-sensitized solar cells. *J. Colloid Interface Sci.,* **2020**, *578*, 184-194.
[http://dx.doi.org/10.1016/j.jcis.2020.04.089] [PMID: 32526522]

[19]　Bhat, V.S.; Kanagavalli, P.; Sriram, G.; B, R.P.; John, N.S.; Veerapandian, M.; Kurkuri, M.; Hegde, G. Low cost, catalyst free, high performance supercapacitors based on porous nano carbon derived from agriculture waste. *J. Energy Storage,* **2020**, *32*, 101829.
[http://dx.doi.org/10.1016/j.est.2020.101829]

[20]　Bhat, V.S.; S, S.; Hegde, G. Review : Biomass derived carbon materials for electrochemical sensors. *J. Electrochem. Soc.,* **2020**, *167*(3), 037526.
[http://dx.doi.org/10.1149/2.0262003JES]

[21]　Karahan, H.E.; Ji, M.; Pinilla, J.L.; Han, X.; Mohamed, A.; Wang, L.; Wang, Y.; Zhai, S.; Montoya, A.; Beyenal, H.; Chen, Y. Biomass-derived nanocarbon materials for biological applications: Challenges and prospects. *J. Mater. Chem. B Mater. Biol. Med.,* **2020**, *8*(42), 9668-9678.
[http://dx.doi.org/10.1039/D0TB01027H] [PMID: 33000843]

[22]　Xuan, J.; Leung, M.K.H.; Leung, D.Y.C.; Ni, M. A review of biomass-derived fuel processors for fuel cell systems. *Renew. Sustain. Energy Rev.,* **2009**, *13*(6-7), 1301-1313.
[http://dx.doi.org/10.1016/j.rser.2008.09.027]

[23]　Sha, T.; Li, X.; Liu, J.; Sun, M.; Wang, N.; Bo, X.; Guo, Y.; Hu, Z.; Zhou, M. Biomass waste derived carbon nanoballs aggregation networks-based aerogels as electrode material for electrochemical sensing. *Sens. Actuators B Chem.,* **2018**, *277*, 195-204.
[http://dx.doi.org/10.1016/j.snb.2018.09.011]

[24]　Pourhosseini, S.E.M.; Norouzi, O.; Salimi, P.; Naderi, H.R. Synthesis of a novel interconnected 3D pore network algal biochar constituting iron nanoparticles derived from a harmful marine biomass as high-performance asymmetric supercapacitor electrodes. *ACS Sustain. Chem.& Eng.,* **2018**, *6*(4), 4746-4758.
[http://dx.doi.org/10.1021/acssuschemeng.7b03871]

[25]　Szczęśniak, B.; Phuriragpitikhon, J.; Choma, J.; Jaroniec, M. Recent advances in the development and applications of biomass-derived carbons with uniform porosity. *J. Mater. Chem. A Mater. Energy Sustain.,* **2020**, *8*(36), 18464-18491.
[http://dx.doi.org/10.1039/D0TA05094F]

[26]　Rudra, S.; K, J.; Thamizharasan, G.; Pradhan, M.; Rani, B.; Sahu, N.K.; Nayak, A.K. Fabrication of Mn3O4-WO3 nanoparticles based nanocomposites symmetric supercapacitor device for enhanced

energy storage performance under neutral electrolyte. *Electrochim. Acta,* **2022**, *406*, 139870.
[http://dx.doi.org/10.1016/j.electacta.2022.139870]

[27]   Gong, Y.; Li, D.; Fu, Q.; Zhang, Y.; Pan, C. Nitrogen self-doped porous carbon for high-performance supercapacitors. *ACS Appl. Energy Mater.,* **2020**, *3*(2), 1585-1592.
[http://dx.doi.org/10.1021/acsaem.9b02077]

[28]   Li, R.; Huang, J.; Li, J.; Cao, L.; Zhong, X.; Yu, A.; Lu, G. Nitrogen-doped porous hard carbons derived from shaddock peel for high-capacity lithium-ion battery anodes. *J. Electroanal. Chem.,* **2020**, *862*, 114044.
[http://dx.doi.org/10.1016/j.jelechem.2020.114044]

[29]   Rodrigues, A.C.; da Silva, E.L.; Oliveira, A.P.S.; Matsushima, J.T.; Cuña, A.; Marcuzzo, J.S.; Gonçalves, E.S.; Baldan, M.R. High-performance supercapacitor electrode based on activated carbon fiber felt/iron oxides. *Mater. Today Commun.,* **2019**, *21*, 100553.
[http://dx.doi.org/10.1016/j.mtcomm.2019.100553]

[30]   Si, C. The development of lignocellulosic biomass in medicinal applications. *Curr. Med. Chem.,* **2019**, *26*(14), 2408-2409.
[http://dx.doi.org/10.2174/092986732614190724160641] [PMID: 31453775]

[31]   Rangaraj, S.; Venkatachalam, R. A lucrative chemical processing of bamboo leaf biomass to synthesize biocompatible amorphous silica nanoparticles of biomedical importance. *Appl. Nanosci.,* **2017**, *7*(5), 145-153.
[http://dx.doi.org/10.1007/s13204-017-0557-z]

[32]   Kalabegishvili, T.; Murusidze, I.; Kirkesali, E.; Rcheulishvili, A.; Ginturi, E.; Kuchava, N.; Bagdavadze, N.; Gelagutashvili, E.; Frontasyeva, M.V.; Zinicovscaia, I.; Pavlov, S.S.; Dmitriev, A.Y. Gold and silver nanoparticles in spirulina platensis biomass for medical application. *Ecol. Chem. Eng. S,* **2013**, *20*(4), 621-631.
[http://dx.doi.org/10.2478/eces-2013-0043]

[33]   Mason, P.E.; Darvell, L.I.; Jones, J.M.; Williams, A. Comparative study of the thermal conductivity of solid biomass fuels. *Energy Fuels,* **2016**, *30*(3), 2158-2163.
[http://dx.doi.org/10.1021/acs.energyfuels.5b02261] [PMID: 27041819]

[34]   Pownraj, C.; Preparation, A. V. A. Preparation and characterization of low cost eco-friendly GAO grafted bio□carbon nanoparticle additive for enhancing the lubricant performance. *Diamond Related Materials,* **2020**, *108*, 107921.
[http://dx.doi.org/10.1016/j.diamond.2020.107921]

[35]   Huang, H.L.; Huang, Z.H.; Chu, Y.C.; Lin, H.P.; Chang, Y.J. Application of metallic nanoparticle-biochars with ionic liquids for thermal transfer fluids. *Chemosphere,* **2020**, *250*, 126219.
[http://dx.doi.org/10.1016/j.chemosphere.2020.126219] [PMID: 32105856]

[36]   Briscoe, J.; Marinovic, A.; Sevilla, M.; Dunn, S.; Titirici, M. Biomass-derived carbon quantum dot sensitizers for solid-state nanostructured solar cells. *Angew. Chem. Int. Ed.,* **2015**, *54*(15), 4463-4468.
[http://dx.doi.org/10.1002/anie.201409290] [PMID: 25704873]

[37]   Youssef, A.M. Polymer nanocomposites as a new trend for packaging applications. *Polym. Plast. Technol. Eng.,* **2013**, *52*(7), 635-660.
[http://dx.doi.org/10.1080/03602559.2012.762673]

[38]   Crosby, A.J.; Lee, J.Y. Polymer nanocomposites: The "Nano" effect on mechanical properties. *Polym. Rev.,* **2007**, *47*(2), 217-229.
[http://dx.doi.org/10.1080/15583720701271278]

[39]   Monteiro, M.S.S.B.; Rodrigues, C.L.; Neto, R.P.C.; Tavares, M.I.B. The structure of polycaprolactone-clay nanocomposites investigated by 1H NMR relaxometry. *J. Nanosci. Nanotechnol.,* **2012**, *12*(9), 7307-7313.
[http://dx.doi.org/10.1166/jnn.2012.6431] [PMID: 23035469]

[40] Valentim, A.C.S.; Tavares, M.I.B.; da Silva, E.O. The effect of the Nb_2O_5 dispersion on ethylene vinyl acetate to obtain ethylene vinyl acetate/Nb_2O_5 nanostructured materials. *J. Nanosci. Nanotechnol.,* **2013**, *13*(6), 4427-4432.
[http://dx.doi.org/10.1166/jnn.2013.7162] [PMID: 23862516]

[41] Monteiro, M.S.S.B.; Cucinelli Neto, R.P.; Santos, I.C.S.; Silva, E.O.; Tavares, M.I.B. Inorganic-organic hybrids based on poly (ε-Caprolactone) and silica oxide and characterization by relaxometry applying low-field NMR. *Mater. Res.,* **2012**, *15*(6), 825-832.
[http://dx.doi.org/10.1590/S1516-14392012005000121]

[42] Zayan, S.; Elshazly, A.; Elkady, M. *In Situ* polymerization of polypyrrole @ aluminum fumarate metal–organic framework hybrid nanocomposites for the application of wastewater treatment. *Polymers,* **2020**, *12*(8), 1764.
[http://dx.doi.org/10.3390/polym12081764] [PMID: 32784539]

[43] Shukla, V. Review of electromagnetic interference shielding materials fabricated by iron ingredients. *Nanoscale Adv.,* **2019**, *1*(5), 1640-1671.
[http://dx.doi.org/10.1039/C9NA00108E] [PMID: 36134227]

[44] Soares, I.L.; Chimanowsky, J.P.; Luetkmeyer, L.; Silva, E.O.; Souza, D.H.S.; Tavares, M.I.B. Evaluation of the influence of modified TiO_2 particles on polypropylene composites. *J. Nanosci. Nanotechnol.,* **2015**, *15*(8), 5723-5732.
[http://dx.doi.org/10.1166/jnn.2015.10041] [PMID: 26369145]

[45] Roy, S.; Singha, N. Polymeric nanocomposite membranes for next generation pervaporation process: Strategies, challenges and future prospects. *Membranes,* **2017**, *7*(3), 53.
[http://dx.doi.org/10.3390/membranes7030053] [PMID: 28885591]

[46] Tambe, S.; Jain, D.; Agarwal, Y.; Amin, P. Hot-melt extrusion: Highlighting recent advances in pharmaceutical applications. *J. Drug Deliv. Sci. Technol.,* **2021**, *63*, 102452.
[http://dx.doi.org/10.1016/j.jddst.2021.102452]

[47] Adnan, M.; Dalod, A.; Balci, M.; Glaum, J.; Einarsrud, M.A. *In Situ* synthesis of hybrid inorganic–polymer nanocomposites. *Polymers,* **2018**, *10*(10), 1129.
[http://dx.doi.org/10.3390/polym10101129] [PMID: 30961054]

[48] Zhang, M.; Li, Y.; Su, Z.; Wei, G. Recent advances in the synthesis and applications of graphene–polymer nanocomposites. *Polym. Chem.,* **2015**, *6*(34), 6107-6124.
[http://dx.doi.org/10.1039/C5PY00777A]

[49] Yadav, R.; Naebe, M.; Wang, X.; Kandasubramanian, B. Structural and thermal stability of polycarbonate decorated fumed silica nanocomposite *via* thermomechanical analysis and *in-situ* temperature assisted SAXS. *Sci. Rep.,* **2017**, *7*(1), 7706.
[http://dx.doi.org/10.1038/s41598-017-08122-7] [PMID: 28794493]

[50] Kayalvizhi Nangai, E.; Saravanan, S. Synthesis, fabrication and testing of polymer nanocomposites: A review. *Mater. Today Proc.,* **2023**, *81*, 91-97.
[http://dx.doi.org/10.1016/j.matpr.2021.02.261]

[51] Sinha Ray, S. *Techniques for characterizing the structure and properties of polymer nanocomposites.Environmentally Friendly Polymer Nanocomposites*; Sinha Ray, S., Ed.; Woodhead Publishing, **2013**, pp. 74-88.

[52] Rizal, S.; Mistar, E.M.; Rahman, A.A.; H P S, A.K.; Oyekanmi, A.A.; Olaiya, N.G.; Abdullah, C.K.; Alfatah, T. Bionanocarbon functional material characterisation and enhancement properties in nonwoven kenaf fibre nanocomposites. *Polymers,* **2021**, *13*(14), 2303.
[http://dx.doi.org/10.3390/polym13142303] [PMID: 34301059]

[53] Chandramohan, D.; Murali, B.; Vasantha-Srinivasan, P.; Dinesh Kumar, S. Mechanical, moisture absorption, and abrasion resistance properties of bamboo–jute–glass fiber composites. *J. Bio Tribocorros.,* **2019**, *5*(3), 66.

[http://dx.doi.org/10.1007/s40735-019-0259-z]

[54] Ceballos, F.; Zhao, H. Ultrafast laser spectroscopy of two-dimensional materials beyond graphene. *Adv. Funct. Mater.,* **2017,** *27*(19), 1604509.
[http://dx.doi.org/10.1002/adfm.201604509]

[55] Fang, Z.; Xing, Q.; Fernandez, D.; Zhang, X.; Yu, G. A mini review on two-dimensional nanomaterial assembly. *Nano Res.,* **2020,** *13*(5), 1179-1190.
[http://dx.doi.org/10.1007/s12274-019-2559-5]

[56] Chang, B.P.; Rodriguez-Uribe, A.; Mohanty, A.K.; Misra, M. A comprehensive review of renewable and sustainable biosourced carbon through pyrolysis in biocomposites uses: Current development and future opportunity. *Renew. Sustain. Energy Rev.,* **2021,** *152*(111666), 111666.
[http://dx.doi.org/10.1016/j.rser.2021.111666]

[57] Kinloch, I.A.; Suhr, J.; Lou, J.; Young, R.J.; Ajayan, P.M. Composites with carbon nanotubes and graphene: An outlook. *Science,* **2018,** *362*(6414), 547-553.
[http://dx.doi.org/10.1126/science.aat7439] [PMID: 30385571]

[58] Mohanty, A.; Misra, M.; Rodriguez-Uribe, A.; Vivekanandhan, S. Hybrid sustainable composites and methods of making and using thereof. US20160229997A1, 2017.

[59] Thippeswamy, B.H.; Maligi, A.S.; Hegde, G. Roadmap of effects of biowaste-synthesized carbon nanomaterials on carbon nano-reinforced composites. *Catalysts,* **2021,** *11*(12), 1485.
[http://dx.doi.org/10.3390/catal11121485]

[60] Bigg, D.M. Mechanical properties of particulate filled polymers. *Polym. Compos.,* **1987,** *8*(2), 115-122.
[http://dx.doi.org/10.1002/pc.750080208]

[61] Agunsoye, J.O.; Odumosu, A.K.; Dada, O. Novel epoxy-carbonized coconut shell nanoparticles composites for car bumper application. *Int. J. Adv. Manuf. Technol.,* **2019,** *102*(1-4), 893-899.
[http://dx.doi.org/10.1007/s00170-018-3206-0]

[62] Bartoli, M.; Giorcelli, M.; Rosso, C.; Rovere, M.; Jagdale, P.; Tagliaferro, A. Influence of commercial biochar fillers on brittleness/ductility of epoxy resin composites. *Appl. Sci.,* **2019,** *9*(15), 3109.
[http://dx.doi.org/10.3390/app9153109]

[63] Matykiewicz, D. Biochar as an effective filler of carbon fiber reinforced bio-epoxy composites. *Processes,* **2020,** *8*(6), 724.
[http://dx.doi.org/10.3390/pr8060724]

[64] Kishore Chowdari, G.; Krishna Prasad, D.V.V.; Devireddy, S.B.R. Physical and thermal behaviour of areca and coconut shell powder reinforced epoxy composites. *Mater. Today Proc.,* **2020,** *26*, 1402-1405.
[http://dx.doi.org/10.1016/j.matpr.2020.02.282]

[65] Langhorst, A.; Paxton, W.; Bollin, S.; Frantz, D.; Burkholder, J.; Kiziltas, A.; Mielewski, D. Heat-treated blue agave fiber composites. *Compos., Part B Eng.,* **2019,** *165*, 712-724.
[http://dx.doi.org/10.1016/j.compositesb.2019.02.035]

[66] Patnaik, P. K.; Mishra, S. K.; Swain, P.T.R. Effect of groundnut shell particulate content on physical and mechanical behavior of jute–epoxy hybrid composite. *J. Instit. Eng.,* **2020,** *103*(1), 65-72.

[67] Parcheta, P.; Datta, J. Environmental impact and industrial development of biorenewable resources for polyurethanes. *Crit. Rev. Environ. Sci. Technol.,* **2017,** *47*(20), 1986-2016.
[http://dx.doi.org/10.1080/10643389.2017.1400861]

[68] Luo, X.; Mohanty, A.; Misra, M. Lignin as a reactive reinforcing filler for water-blown rigid biofoam composites from soy oil-based polyurethane. *Ind. Crops Prod.,* **2013,** *47*, 13-19.
[http://dx.doi.org/10.1016/j.indcrop.2013.01.040]

[69] Seto, C.; Chang, B.P.; Tzoganakis, C.; Mekonnen, T.H. Lignin derived nano-biocarbon and its

deposition on polyurethane foam for wastewater dye adsorption. *Int. J. Biol. Macromol.,* **2021**, *185*, 629-643.
[http://dx.doi.org/10.1016/j.ijbiomac.2021.06.185] [PMID: 34216664]

[70]   Snowdon, M.R.; Wu, F.; Mohanty, A.K.; Misra, M. Comparative study of the extrinsic properties of poly(lactic acid)-based biocomposites filled with talc *versus* sustainable biocarbon. *RSC Advances,* **2019**, *9*(12), 6752-6761.
[http://dx.doi.org/10.1039/C9RA00034H] [PMID: 35518480]

[71]   Harris, A.M.; Lee, E.C. Heat and humidity performance of injection molded PLA for durable applications. *J. Appl. Polym. Sci.,* **2010**, *115*(3), 1380-1389.
[http://dx.doi.org/10.1002/app.30815]

[72]   Najafi, N.; Heuzey, M.C.; Carreau, P.J. Polylactide (PLA)-clay nanocomposites prepared by melt compounding in the presence of a chain extender. *Compos. Sci. Technol.,* **2012**, *72*(5), 608-615.
[http://dx.doi.org/10.1016/j.compscitech.2012.01.005]

[73]   Armentano, I.; Bitinis, N.; Fortunati, E.; Mattioli, S.; Rescignano, N.; Verdejo, R.; Lopez-Manchado, M.A.; Kenny, J.M. Multifunctional nanostructured PLA materials for packaging and tissue engineering. *Prog. Polym. Sci.,* **2013**, *38*(10-11), 1720-1747.
[http://dx.doi.org/10.1016/j.progpolymsci.2013.05.010]

[74]   Huang, J.W.; Chang Hung, Y.; Wen, Y.L.; Kang, C.C.; Yeh, M.Y. Polylactide/nano- and micro-scale silica composite films. II. Melting behavior and cold crystallization. *J. Appl. Polym. Sci.,* **2009**, *112*(5), 3149-3156.
[http://dx.doi.org/10.1002/app.29699]

[75]   Arrigo, R.; Bartoli, M.; Malucelli, G. Poly(lactic Acid)–biochar biocomposites: Effect of processing and filler content on rheological, thermal, and mechanical properties. *Polymers,* **2020**, *12*(4), 892.
[http://dx.doi.org/10.3390/polym12040892] [PMID: 32290601]

[76]   Jasim, S.M.; Ali, N.A. Properties characterization of plasticized polylactic acid /biochar (bio carbon) nano-composites for antistatic packaging. *Iraqi J. Phys.,* **2019**, *17*(42), 13-26.
[http://dx.doi.org/10.30723/ijp.v17i42.419]

[77]   Xu, H.; Xie, L.; Li, J.; Hakkarainen, M. Coffee grounds to multifunctional quantum dots: Extreme nanoenhancers of polymer biocomposites. *ACS Appl. Mater. Interfaces,* **2017**, *9*(33), 27972-27983.
[http://dx.doi.org/10.1021/acsami.7b09401] [PMID: 28770986]

[78]   Ho, M.; Lau, K.; Wang, H.; Hui, D. Improvement on the properties of polylactic acid (PLA) using bamboo charcoal particles. *Compos., Part B Eng.,* **2015**, *81*, 14-25.
[http://dx.doi.org/10.1016/j.compositesb.2015.05.048]

[79]   Adeosun, S.; Taiwo, O.; Akpan, E.; Gbenebor, O.; Gbagba, S.; Olaleye, S. Mechanical characteristics of groundnut shell particle reinforced polylactide nano fibre. *Materia,* **2016**, *21*(2), 482-491.
[http://dx.doi.org/10.1590/S1517-707620160002.0045]

[80]   Wang, L.; Tong, Z.; Ingram, L.O.; Cheng, Q.; Matthews, S. Green composites of poly (Lactic Acid) and sugarcane bagasse residues from bio-refinery processes. *J. Polym. Environ.,* **2013**, *21*(3), 780-788.
[http://dx.doi.org/10.1007/s10924-013-0601-3]

[81]   Nezakati, T.; Seifalian, A.; Tan, A.; Seifalian, A.M. Conductive polymers: Opportunities and challenges in biomedical applications. *Chem. Rev.,* **2018**, *118*(14), 6766-6843.
[http://dx.doi.org/10.1021/acs.chemrev.6b00275] [PMID: 29969244]

[82]   Gabe, A.; Mostazo-López, M.J.; Salinas-Torres, D.; Morallón, E.; Cazorla-Amorós, D. *Synthesis of conducting polymer/carbon material composites and their application in electrical energy storage. Hybrid Polymer Composite Materials*; Elsevier, **2017**, pp. 173-209.

[83]   Hu, Y.; Tong, X.; Zhuo, H.; Zhong, L.; Peng, X. Biomass-based porous N-self-doped carbon framework/polyaniline composite with outstanding supercapacitance. *ACS Sustain. Chem.& Eng.,* **2017**, *5*(10), 8663-8674.

[http://dx.doi.org/10.1021/acssuschemeng.7b01380]

[84] Yu, M.; Han, Y.; Li, Y.; Li, J.; Wang, L. Polypyrrole-anchored cattail biomass-derived carbon aerogels for high performance binder-free supercapacitors. *Carbohydr. Polym.,* **2018**, *199*, 555-562.
[http://dx.doi.org/10.1016/j.carbpol.2018.04.058] [PMID: 30143162]

[85] Das, O.; Bhattacharyya, D.; Hui, D.; Lau, K.T. Mechanical and flammability characterisations of biochar/polypropylene biocomposites. *Compos., Part B Eng.,* **2016**, *106*, 120-128.
[http://dx.doi.org/10.1016/j.compositesb.2016.09.020]

[86] Wan, C.; Li, J. Wood-derived biochar supported polypyrrole nanoparticles as a free-standing supercapacitor electrode. *RSC Advances,* **2016**, *6*(89), 86006-86011.
[http://dx.doi.org/10.1039/C6RA17044G]

[87] Das, C.; Tamrakar, S.; Mielewski, D.; Kiziltas, A.; Xie, X. Newly developed biocarbon to increase electrical conductivity in sustainable polyamide 12 composites. *Polym. Compos.,* **2022**, *43*(11), 8084-8094.
[http://dx.doi.org/10.1002/pc.26953]

[88] Yang, Q.; Yang, W.; Shi, Y.; Yu, L.; Li, X.; Yu, L.; Dong, Y.; Zhu, Y.; Fu, Y. Aligned polyaniline/porous biomass carbon composites with superior microwave absorption properties. *J. Mater. Sci. Mater. Electron.,* **2019**, *30*(2), 1374-1382.
[http://dx.doi.org/10.1007/s10854-018-0407-0]

[89] Peterson, S.C.; Kim, S. Using heat-treated starch to modify the surface of biochar and improve the tensile properties of biochar-filled styrene–butadiene rubber composites. *J. Elastomers Plast.,* **2019**, *51*(1), 26-35.
[http://dx.doi.org/10.1177/0095244318768636]

[90] Selvaraj, V.; Raghavarshini, T.R.; Alagar, M. Design and development of bio-carbon reinforced hetero structured biophenolics polybenzoxazine-epoxy hybrid composites for high performance applications. *J. Polym. Res.,* **2021**, *28*(5), 174.
[http://dx.doi.org/10.1007/s10965-020-02338-4]

[91] Rhim, J.W.; Park, H.M.; Ha, C.S. Bio-nanocomposites for food packaging applications. *Prog. Polym. Sci.,* **2013**, *38*(10-11), 1629-1652.
[http://dx.doi.org/10.1016/j.progpolymsci.2013.05.008]

[92] Umerah, C.O.; Kodali, D.; Head, S.; Jeelani, S.; Rangari, V.K. Synthesis of carbon from waste coconutshell and their application as filler in bioplast polymer filaments for 3D printing. *Compos., Part B Eng.,* **2020**, *202*, 108428.
[http://dx.doi.org/10.1016/j.compositesb.2020.108428]

[93] Kanmani, P.; Rhim, J-W. *Nano and nanocomposite antimicrobial materials for food packaging applications.Progress in Nanomaterials for Food Packaging*; Future Science Ltd, **2014**, pp. 34-48.
[http://dx.doi.org/10.4155/ebo.13.303]

[94] Vazhayal, L.; Wilson, P.; Prabhakaran, K. Waste to wealth: Lightweight, mechanically strong and conductive carbon aerogels from waste tissue paper for electromagnetic shielding and CO_2 adsorption. *Chem. Eng. J.,* **2020**, *381*, 122628.
[http://dx.doi.org/10.1016/j.cej.2019.122628]

[95] Devi, G.; Nagabhooshanam, N.; Chokkalingam, M.; Sahu, S.K. EMI shielding of cobalt, red onion husk biochar and carbon short fiber- PVA composite on X and Ku band frequencies. *Polym. Compos.,* **2022**, *43*(9), 5996-6003.
[http://dx.doi.org/10.1002/pc.26898]

[96] Daniel, S.; Joseph, S.; Alapat, P.S.; Kalarikkal, N.; Thomas, S. Silver-sandwiched natural rubber/st-LDH/MWCNT hybrid bio-nano-composite system as a high-performing multimedia laminated electromagnetic interference shield through a tripling mechanism. *ACS Sustain. Chem.& Eng.,* **2022**, *10*(45), 14897-14913.
[http://dx.doi.org/10.1021/acssuschemeng.2c04820]

CHAPTER 2

Biochar-thermoplastic Polymer Composites: Recent Advances and Perspectives

Giulio Malucelli[1,2,*]

[1] *Department of Applied Science and Technology, Politecnico di Torino, Viale Teresa Michel 5, 15129, Alessandria, Italy*

[2] *Consorzio Interuniversitario Nazionale per la Scienza e Tecnologia dei Materiali (INSTM), Via G. Giusti 9, 50121 Firenze, Italy*

Abstract: To fulfill the current circular economy concept, several attempts to reuse and valorize wastes and by-products coming from different sectors (such as the agri-food, textile, and packaging industries, among others) are being carried out at least at a lab scale by academics, despite the increasing interest that also involves the industrial world. One of the up-to-date strategies to transform wastes and by-products into new added-value systems refers to the production of biochar (BC), a carbonaceous solid residue derived from the thermo-chemical conversion, under controlled conditions, of wastes or, more generally, biomasses. Apart from its conventional uses (such as for soil remediation, heat and power production, low-cost carbon sequestration, and as a natural adsorbent, among others), BC is gaining a continuously increasing interest as a multifunctional micro-filler for different thermoplastic and thermosetting polymer matrices. Undoubtedly, the wide possibility of producing BC from different biomass sources, wastes, and by-products offers an attractive prospect toward a circular bio-economy with "zero waste". When incorporated into a polymer at different loadings, BC can provide thermal and electrical conductivity, EMI shielding features, enhanced mechanical properties, and flame retardance as well. This chapter aims to summarize the current achievements in the design, preparation, and characterization of thermoplastic polymer/biochar composites, discussing the current limitations/ drawbacks, and providing the reader with some perspectives for the future.

Keywords: Biochar, Degree of Graphitization, Electrical Conductivity, EMI Shielding, Flame Retardance, Mechanical Behavior, Thermoplastic-biochar Composites, Thermal Conductivity.

---

[*] **Corresponding author Giulio Malucelli:** Department of Applied Science and Technology, Politecnico di Torino, Viale Teresa Michel 5, 15121 Alessandria, Italy; Tel: +390131229369; E-mail: giulio.malucelli@polito.it

Deepa Kodali & Vijaya Rangari (Eds.)
All rights reserved-© 2023 Bentham Science Publishers

1. INTRODUCTION

Nowadays, both environmental protection and the increasing depletion of fossil resources (crude oil, carbon, and natural gas) are stimulating not only academic research but also the industrial world toward the design and development of sustainable manufacturing strategies [1, 2]. In particular, the current research trends indicate a huge interest in the valorization of low environmental impact products derived from renewable wastes and biomasses for different uses [3] For an effective exploitation, these products should satisfy different requirements: in particular, they should exhibit acceptable performances for the envisaged applications, highlighting their suitability, as sustainable materials, for replacing fossil-derived counterparts. In addition, they should exhibit a certain "flexibility" of use, *i.e.*, providing either multifunctional features or being exploitable for diverse application sectors [4].

In this perspective, particular attention is currently being paid to biochar (BC), a renewable carbonaceous product that can be obtained through the thermo-chemical conversion, under controlled conditions, of different wastes, by-products, and biomasses. Generally speaking, BC is a micron-sized filler that shows a porous structure with an extended surface area and high thermal stability, notwithstanding acceptable thermal and electrical conductivities [5 - 8]. Thanks to its high porosity, biochar has been extensively exploited for filtration purposes, remediation of pollutants [9], and even as an antimicrobial product, taking advantage of the possibility of tuning its final properties based on the feedstock and the conditions adopted for its production [10 - 14]. More specifically, the graphitization degree achieved during BC production is strictly related to the composition of the pristine wastes, by-products, and biomasses, as well as to the adopted pyrolysis temperatures and residence times. Usually, increasing these two parameters accounts for an increased graphitization degree, determining, at the same time, a lowering of the hydrogen and oxygen content [15]. Conversely, when pyrolysis is performed at low temperatures, some volatile residues can be entrapped in the pores, hence decreasing the porosity of the final material [16]. Although its conventional uses in the agriculture sector (namely as a nutrient source, for soil conditioning and for remediation and absorption of contaminants, among others [17, 18]), in the last decade, biochar started to be employed as an electrode material in the design of supercapacitors [19, 20], as a catalyst in the processing of syngas [21], and as an anode for fuel cells [22]. Because of its peculiar features, the ease of tailoring its structure and morphology, and the wide functionality that can be provided, biochar is also currently being exploited for replacing conventional carbon-based materials in the design and manufacturing of polymer composites exhibiting improved thermal, mechanical, and electrical features [23 - 25]. The scientific interest in BC is remarkably rising, as witnessed by the increasing number of articles that appeared in scientific journals (Fig. **1**).

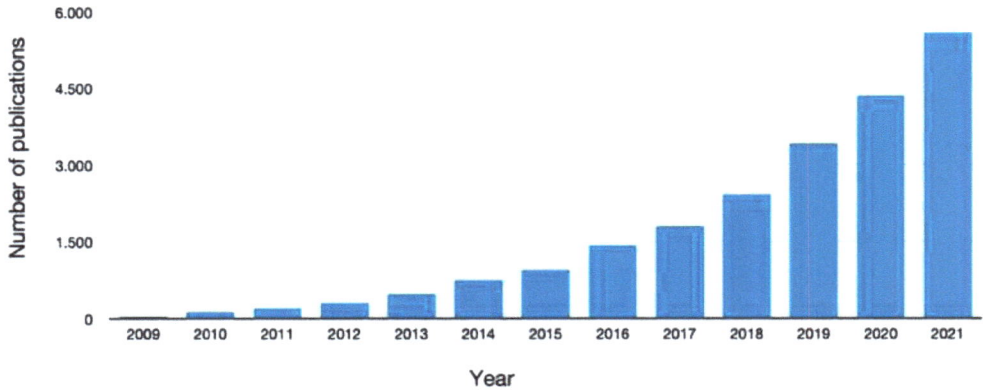

Fig. (1). Number of publications dealing with biochar from 2009 to 2021. (Source: Web of ScienceTM, accessed on 2 December 2022).

In this view, the present chapter reviews the latest progress in the design, preparation, and characterization of thermoplastic polymer composites containing biochar, highlighting the structure-property-production relationships established based on biochar structure, morphology, composition, loading, and possible functionalization. The current limitations are also discussed, providing the reader with some perspectives for the future development of these composite materials.

2. SYNTHESIS OF BIOCHAR (BC)

As mentioned in the previous paragraph, biochar is obtained by thermally treating (*i.e.*, pyrolyzing) biomasses in an O_2-free (or O_2-limited) environment; usually, the thermal treatment is carried out above 350°C for different times. It is possible to process different biomasses/wastes/by-products, such as wood, food wastes, plant residues, industrial biowastes, and animal carcasses, among others [26]. The chemical treatment generally preserves a significant part of the carbon of the biomass (*i.e.*, about one-third), leading to the formation of non-condensable gases and biofuel (*i.e.*, combustible bio-oil) as well [27, 28]. Besides, thanks to the pyrolysis process and the related structural changes, the porosity and surface area of BC significantly increase compared with that of the pristine biomass [29].

The pyrolysis takes place according to three step-wise reactions [30]: first, moisture is removed, then lignocellulosic components of the biomass undergo depolymerization and fragmentation processes, during which bonded water and carbon dioxide are further removed. Finally, re-condensation and re-polymerization reactions occur, thus leading to char formation. Several experimental parameters, including temperature and residence time, pressure, heating rate, and gas environment maintained during the pyrolysis, significantly contribute to determining the final properties of the obtained biochar (*i.e.*, porosity, surface area, carbon structure and degree of graphitization, particle size and size distribution, ash content, among others) [31]. In particular, the adopted heating rate allows for classifying the pyrolysis as *slow* (0.1-10°C/s; residence times between 5 min and several hours; 25-50% biochar yields), *fast* (10-200°C/s; residence times between 10 and 25 min; 15-25% biochar yields), and *flash* (beyond 1000°C/s; residence times below 1 min; 5-15% biochar yields) [32].

Quite recently, the synthesis of biochar has been moving from conventional thermochemical treatments to advanced pyrolytic techniques that usually operate at low temperatures, are cost-effective, and allow for achieving optimized BC features. They comprise hydro/wet pyrolysis, co-pyrolysis, ammonia ambiance pyrolysis, microwave-assisted pyrolysis, catalytic pyrolysis, and steam-assisted pyrolysis. All these advanced strategies are exploited for enhancing the yields, porosity, and surface area, as well as the functionalization degree of the resulting biochars [30, 33, 34].

3. BIOCHAR-THERMOPLASTIC POLYMER COMPOSITES

3.1. Biochar in Polyolefins

Biochar has quite extensively been exploited for the design of thermoplastic composites. As reported in the scientific literature, one of its first uses dates back to 2015, when BC derived from the pyrolysis of landfill pine wood carried out at either 400 or 450°C, was melt compounded with **polypropylene** at different concentrations (within 6 and 30 wt.%), in combination with wood (at 30 wt.% loading) [35]. It was found that the incorporation of 24 wt.% of biochar greatly promoted the adhesion between wood and the polyolefin interface: in particular, BC did not affect both Young's modulus and tensile strength but remarkably increased the flexural behavior of the prepared polypropylene/wood composites. In addition, a remarkable reduction in the elongation at break was observed when the carbonaceous filler exceeded 15 wt.%. Further, the same research group investigated the effect of the waste feedstock on the mechanical behavior of polypropylene composites embedding 24 wt.% of biochar (obtained from six different wastes) and 30 wt.% of wood [36]. The incorporation of BC improved the flexural and tensile moduli with respect to the composites containing wood filler only: more specifically, the mechanical behavior was enhanced by increasing BC content and its surface area.

Then, the fire behavior of polypropylene-biochar (from landfill pine wood waste pyrolyzed at 500°C and subsequently activated at 900 °C; maximum loading: 30 wt.%) composites was investigated. As revealed by cone calorimetry tests, increased filler loadings strongly lowered the peak of heat release rate (by -54% for the composite containing 35 wt.% of BC) with respect to the unfilled polyolefin, hence highlighting the suitability of biochar for limiting the oxygen transfer to the polyolefin during the combustion process [37] In parallel, thanks to mechanical interlocking due to the infiltration of the melted macromolecular chains into the porous structure of the biochar [38], both flexural and tensile moduli showed a monotonic increase with increasing BC content.

Remarkably improved flame retardant properties were achieved also when biochar was incorporated together with ammonium polyphosphate [39] into wood fiber/ polypropylene composites, although with a negative effect on the mechanical behavior, due to the reduction of the mechanical concatenation in the matrix between biochar and the polymer, as the porous filler structure was partially occupied by the flame retardant particles. A similar fire and mechanical behavior was observed when the recipe of these composites was further modified, introducing $Mg(OH)_2$ [40].

Quite recently, Paleri and co-workers [41] demonstrated that the experimental condition adopted for the pyrolysis process can remarkably affect the mechanical behavior of biochar-PP composites (filler content: 30 wt.%). To this aim, distillers' dried grains with solubles (coming from the corn ethanol process) were employed as a starting material for obtaining BC particles differing as far as the pyrolysis temperature is considered. The obtained biochar showed increased ash content, as well as lowered functionality when it was pyrolyzed at high temperatures (*i.e.*, 1000°C). In particular, 700°C was found to be a good compromise for balancing stiffness and toughness in the resulting composites that exhibited increased flexural modulus (by about 55%), impact strength (+27%), Young's modulus (+40%) with respect to neat polypropylene. In addition, as assessed by rheological analyses, the composites containing BC pyrolyzed at 700°C showed higher values of storage modulus and complex viscosity with respect to the other composites containing biochar obtained at different pyrolysis temperatures, hence demonstrating enhanced compatibility with the macromolecular chains of the polyolefin.

The introduction of a coupling agent (namely, a maleic anhydride-derived copolymer) in the formulation of BC-polypropylene composites did not significantly affect their overall mechanical behavior that was, once again, strictly correlated with the pyrolysis temperature [42]. In particular, beyond 400°C, the progressively decreased number of functional groups in BC accounted for the occurrence of weak BC/polyolefin interactions, hence worsening the mechanical performances. Conversely, in these pyrolysis conditions, the obtained composites showed increased hydrophobicity and dimensional stability.

The incorporation of biochar into polypropylene also influences the crystallization behavior of the polymer: in fact, the filler acted as a nucleating agent, hence increasing the crystallinity degree, as assessed by Alghyamah and co-workers [43]. In particular, different biochar particles, obtained from waste biomass at different pyrolysis temperatures (namely, 300, 400, 500, 600, and 700°C), were incorporated into polypropylene at different loadings (from 5 to 20 wt.%). As revealed by DSC analyses (Fig. **2**), the presence of increasing amounts of BC accounted for an increased degree of crystallinity (from 17.6 to 40.8% for neat polypropylene and for the system embedding 20 wt.% of biochar pyrolyzed at 700°C, respectively). Besides, the shape and morphology of the formed PP crystals were strictly correlated with the pyrolysis temperatures that were responsible for the porosity and particle surface area. In particular, disk-shaped PP crystals were obtained at high pyrolysis temperatures (accounting for a higher infiltration of the macromolecular chains into the porous structure of biochar, while rod-like structures were formed at low pyrolysis temperatures.

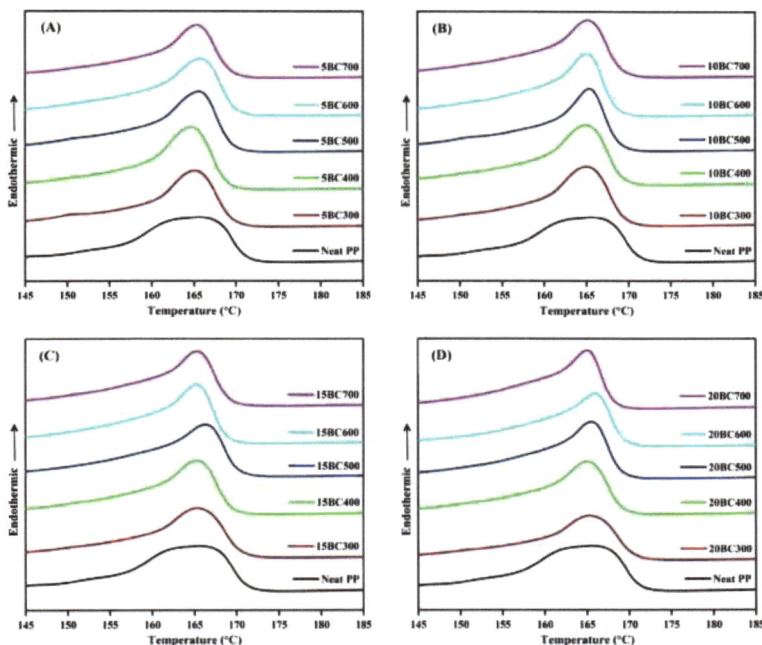

Fig. (2). DSC thermograms of polypropylene (PP) and its composites containing different biochar loadings: (A) 5, (B) 10, (C) 15, and (D) 20 wt.%. BCXXX= biochar particles obtained at XXX°C (pyrolysis temperature). Reprinted from [43] under CC BY 4.0 license.

Among other thermoplastic matrices, BC has been widely utilized for preparing **polyethylene**-based composites. In particular, for functional applications (*i.e.*, for the design of electrically conductive and antistatic materials, and of sensors as well), biochar was dispersed into **Ultra-High Molecular Weight Polyethylene** (UHMWPE), taking advantage of the segregated microstructure generally attained in UHMWPE-based composites. This peculiar morphology allows for efficiently lowering the interfacial electrical resistance between the two components, favoring the formation of a continuous interconnected conductive network [44].

Li and co-workers [45] dispersed biochar derived from bamboo charcoal in UHMWPE (in a mixture with linear low-density polyethylene (LLDPE)), by means of melt extrusion, achieving high filler loadings, up to 80 wt.%. The electrical percolation was reached beyond 60 wt.% of BC: at 80 wt.% filler loading, the electrical conductivity was as high as 107.6 S/m: This finding was ascribed to the effective electron transfer mechanism through direct physical contact or tunneling effects. Besides, the composites containing 80 wt.% of BC showed EMI shielding properties.

A nice example referring to the design and preparation of flexible biochar-UHMWPE composites showing good mechanical features and tailored electrical conductivity was proposed by Liu and co-workers [46], who exploited extrusion and compression molding for obtaining composites filled with three different biochar particles derived from the pyrolysis (carried out at 500, 700, 900, and 1100°C) of charcoal of pine, apple, and bamboo. The adopted processing conditions allowed for good dispersion of the biochars in the polyolefin; besides, irrespective of the type of BC, the obtained composites were flexible even in the presence of 70 wt.% of the filler. It is noteworthy that for these composites embedding the biochar particles obtained at 1100°C, the electrical conductivity was as high as 3.9×10^{-1} (bamboo charcoal), 3.7×10^{-1} (apple charcoal), and 3.0×10^{-1} S/cm (pine charcoal). These findings were ascribed to the creation of biochar conductive networks in the polyolefin, which favored the electron transfer between adjacent filler particles.

Then, it was verified that it is possible to achieve low electrical percolation threshold values (as low as 2 vol.%) even in biochar-UHMWPE composites that exhibit a fully developed segregated structure [47]. For this purpose, high-shear mixing with hot compression techniques were combined, in order to distribute the biochar particles (resulting from the pyrolysis of nano bamboo charcoal performed in a tubular furnace at 1000 °C for 1 h) only at the interface between adjacent UHMWPE particles, hence ensuring the formation of a segregated electrically conductive biochar network. Further, electrical conductivity as high as 1.1×10^{-2} S/cm was measured in the composites incorporating 7 vol.% of biochar.

Finally, the designed composite systems showed increased tensile strength (that reached 30.8 MPa for the composites embedding 3 wt.% of biochar) and thermal stability with increasing the BC content.

Apart from the aforementioned applications, biochar-UHMWPE composites containing bamboo charcoal pyrolyzed at two different temperatures (namely, 800 and 1000°C) were recently proposed for biomedical (orthopedic) uses [48]. In particular, the pyrolysis temperature was found to play a key role in providing either enhanced hardness and increased Young's modulus (when biochar was obtained at 1000°C), or improved tensile strength, wettability, and biocompatibility, as well as decreased friction coefficient (when biochar was synthesized at low temperature, *i.e.*, 800°C).

Recently, the packaging and automotive fields have benefited from the use of new composites made of biochar and **high-density polyethylene** (HDPE). As an example, four different biochars obtained from agricultural wastes (*i.e.*, rice husk) were successfully incorporated into HDPE at 50 wt.% loading by means of compounding and subsequent injection molding [49]. As compared with the unfilled polyolefin, the resulting composites exhibited enhanced mechanical properties (particularly referring to tensile strength and resistance to stress relaxation and creep), improved thermal stability, and ameliorated fire behavior: the trend of these features increased with increasing the filler loading. Similar to what was described for UHMWPE-biochar composites, the pyrolysis temperature played a key role in favoring the interlocking mechanism, *i.e.*, the infiltration of the HDPE macromolecules into the biochar pores: in particular, increasing the pyrolysis temperature accounted for an increase of both porosity and particle surface area, hence leading to an enhanced flexural strength of the resulting composites [50]. This interlocking mechanism was recently interpreted on the basis of the rheological behavior of HDPE-biochar composites by Arrigo and co-workers [51], who incorporated biochar derived from waste brewed coffee powder (pyrolyzed in a tubular furnace at 700 °C for 1 h in N_2 atmosphere) at different loadings (namely, 1, 2.5, 5, and 7.5 wt.%). The observed decrease in the relaxation dynamics of HDPE macromolecular chains (Fig. 3) was attributed to the confinement on the filler surface or its pores. Besides, in the non-linear regime (Fig. 4), the composites containing more than 2.5 wt.% of biochar showed a feeble strain overshoot behavior that suggested the occurrence of interactions between the two phases, able to delay the mobility of the polyolefin macromolecules, impeding their full relaxation.

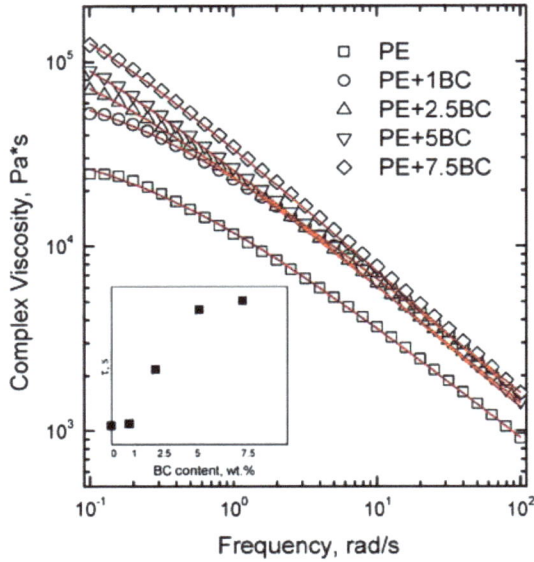

Fig. (3). Complex viscosity as a function of frequency of unfilled HDPE and its composites (continuous lines correspond to cross-model fit). In the inset, the variation of the relaxation time (τ) *vs.* BC content of is reported. XBC= wt.% biochar loading. Reprinted from [51] under CC BY 4.0 license.

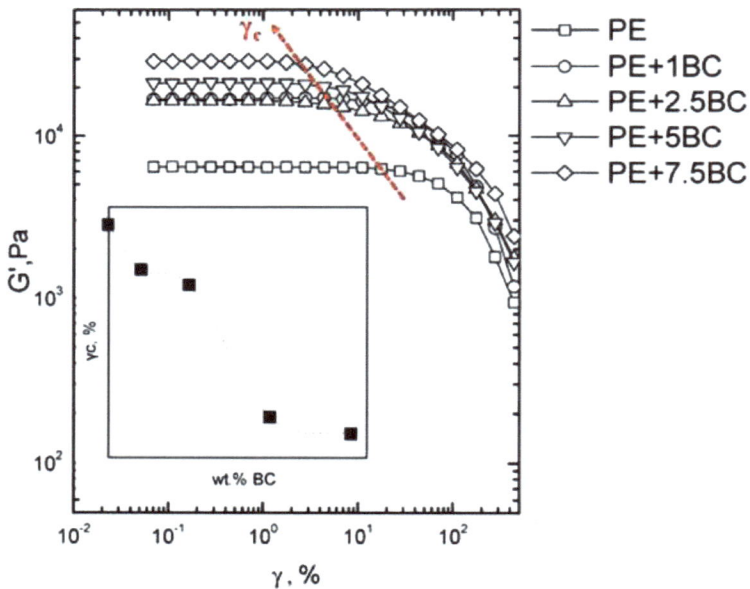

Fig. (4). Complex viscosity as a function of the frequency of unfilled HDPE and its composites (continuous lines correspond to cross-model fit). In the inset, the variation of the relaxation time (τ) *vs.* BC content of is reported. xBC= wt.% biochar loading. Reprinted from [51] under CC BY 4.0 license.

3.2. Biochar in Polyamides

The scientific literature also describes some well-written papers on **polyamide**-biochar composites showing an overall enhanced mechanical behavior; polyamide 6 (PA6) and polyamide 6,10 (PA6,10) are the most preferred polymer matrices. Concerning the latter, biochar derived from the pyrolysis of miscanthus fibers (performed at 500°C) was incorporated at 20 wt.% loading into PA6,10 by means of a micro-compounder and then subjected to injection molding [52]. The effect of the BC particle size, shape, and size distribution (after crashing, milling, and possibly sieving processes) on the thermal and mechanical performances of the resulting composites was thoroughly studied. The impact resistance of the composites enhanced with decreasing the particle size; further, maximum values for flexural and tensile strengths were achieved when the biochar was milled and unsieved. Finally, the biochar accounted for slowing down the crystallization rate: this finding was ascribed to the restricted mobility of the macromolecular chain exerted by the rigid filler.

In a further research effort [53], the effect of the pyrolysis temperature on the mechanical behavior of PA6-biochar composites was examined. For this purpose, miscanthus fibers were thermally-treated either at 500 or at 900°C. Thanks to a better polymer/biochar interfacial adhesion ascribed to different surface functionalities of the latter, the composites containing biochar (20 wt.% loading) processed at 500°C exhibited higher flexural and tensile strength (about +31 and +20%, respectively) as compared with the counterparts embedding the biochar obtained at 900°C.

The scientific literature also documents some successful examples of design, preparation, and characterization of bio-sourced polyamides containing biochar: in this context, Watt and co-workers [54] compounded PA4,10 (a 76% bio-sourced polyamide) with biochars (ball-milled and then sieved to get particles below 300 μm) obtained from corn cob pyrolyzed at 350, 500, and 900°C, using a micro-compounder. As the pyrolysis temperature increased, biochar exhibited higher disordered structures with higher graphitization degrees. Besides, the filler produced at 350°C accounted for enhanced rheological behavior due to particle encapsulation effects at the polyamide interface, without significantly affecting the overall mechanical behavior.

3.3. Biochar in Polyesters

Polyesters represent another important polymer family that has been widely investigated in combination with biochar. In particular, much attention has been devoted to **poly(lactic acid)** (PLA)-biochar composites. In fact, the incorporation of biochar into this polyester can remarkably overcome the main drawbacks of the biodegradable polymer matrix, which refer to its low thermal stability, high brittleness, and low propensity to crystallize, among others [55].

In this context, Ho *et al.* [56] reported a significant enhancement of the mechanical behavior of PLA embedding biochar particles (from 2.5 to 10 wt.% loadings) obtained from the pyrolysis of bamboo. More specifically, tensile strength, flexural strength, and ductility index showed a monotonic increase with increasing the filler loading; however, beyond 7.5 wt.% of BC (at which tensile strength, flexural strength, and ductility index achieved their maximum values, corresponding to an increase by 43, 95 and 52% as compared to the unfilled polymer, respectively), a detrimental effect on the overall mechanical behavior was observed, due to the occurrence of particle aggregation phenomena.

Conversely, Quian and co-workers demonstrated the possibility of homogeneously dispersing ultrafine (1000 mesh) bamboo biochar into poly(lactic acid) up to 30 wt.% loadings [57], at which Young's modulus, tensile strength, and impact strength reached their maximum values (*i.e.*, about 560 MPa, 14 MPa, and 20.5 J/m, respectively). Further, the ductility decreased with increasing the filler loading and, regardless of this latter, it was always below that of the unfilled polymer matrix.

The scientific literature well documents the possibility of (i) using selected coupling agents to enhance the interface between biochar and poly(lactic acid) or (ii) chemically modifying the biochar or the macromolecular chains, aiming at improving the properties of the resulting composites. As an example, Salak and co-workers [58] incorporated biochar particles derived from kudzu biomass, which were further reacted with phthalic anhydride by means of noncatalytic esterification reactions. The modified biochar at 50 wt.% loading accounted for enhanced compatibility with the polymer matrix, leading to ameliorated physicomechanical features, particularly referring to hardness (around +37 and +3% as compared with unfilled PLA and PLA embedding unmodified biochar, respectively), Young's modulus (about +28 and +10%, respectively), and to water resistance.

Quian and co-workers [59] exploited a surface modification of ultra-fine moso bamboo biochar with solutions of nitric acid and sodium hydroxide at different concentrations, to induce the grafting of amino or carbonyl/carboxyl groups, respectively, onto the carbon filler. The so-obtained modified biochars were incorporated into poly(lactic acid), by means of a co-rotating twin screw mini-extruder. The modification of the biochar strongly affected the mechanical behavior of the prepared composites: in particular, the treatment with 15 wt.% of

NaOH solution accounted for the achievement of maximum Young's modulus and tensile strength (about 1012 and 65 MPa, respectively), while the maximum impact strength (26 kJ/m^2) and elongation at break (15%) were reached treating the biochar with nitric acid solutions at 54 and 39 wt.%, respectively.

Generally speaking, apart from the observed enhanced mechanical behavior, the incorporation of biochar (irrespective of its structure, composition, and morphology) into poly(lactic acid) usually accounted for a worsening of the thermal stability (with a lowering of both T_{onset} and T_{max} values), melting point and glass transition temperature of the biopolymer: this finding, well described in the scientific literature, was ascribed to (i) the presence of K in the biochar, which acted as a catalyst, hence anticipating the polymer degradation [60, 61] or (ii) hydrolytic reactions due to the presence of hydroxyls located on the biochar surface [21, 62].

Ternary blends containing polylactic acid, biochar, and a third constituent have extensively been exploited for further enhancing the mechanical performances of the obtained composites. Sheng and co-workers [63] prepared polylactic acid composites embedding ultra-fine bamboo char (from 0.25 to 4 wt.% loading) and silylated bamboo cellulose nanowhiskers (at 2.5 wt.% loading), using a solution casting method. The obtained composites exhibited improved toughness and mechanical strength, due to their peculiar morphology: more specifically, the reinforcing mechanism exerted by the biochar was explained on the basis of the formation of small cavities around the ultra-fine biochar particles by the polymer during the stretching process, which originated a core-shell dispersion structure that favored the dispersion of the stretching energy in a more effective and uniform way.

Further, the incorporation of graphite in polylactic acid-biochar composites has been found very promising for the development of radiation-sensitive devices and sensors, wearable or portable electronics, and EMI shielding systems: In particular, it was possible to design thin composite films (0.25 mm thickness) showing high electrical conductivity (beyond 30 S/m) and remarkable EMI shielding efficiency (beyond 30 dB) [64].

Apart from polylactic acid, other polyester matrices have been investigated for the preparation of biochar-reinforced composites: they include poly(ethylene terephthalate), poly(butylene terephthalate), and poly(trimethylene terephthalate).

As far as **poly(ethylene terephthalate)** is considered, Snowdon and co-workers [65] exploited a statistical approach for assessing the effect of exploiting recycled poly(ethylene terephthalate) (RPET), biochar (derived from *Miscanthus* pyrolyzed at 650 C and subsequently ball-milled to an average particle size of about 1.7μm), and a chain extender (namely, styrene-acrylic-glycidyl methacrylate - SAGMA) on the mechanical behavior of impact-modified PET biocomposites (toughening agent: poly(ethylene n-butylene-acrylate glycidyl methacrylate - EBAGMA). To this aim, a full factorial design was employed for the mechanical tests (tensile and impact tests) performed on the prepared samples; the results were thoroughly examined, evaluating how the components of the composite recipe and their ratios affected the overall mechanical features. A scheme of the possible interactions taking place among the different components is depicted in Fig. **(5)**. The chain extender was the most effective component in changing the mechanical behavior (*i.e.*, Young's modulus, elongation at break, tensile strength at yield, and impact strength), with biochar being the second most important factor. The optimum composite recipe for achieving balanced mechanical properties comprised up to 25 wt.% of recycled PET, 10 wt.% of biochar, and the chain extender, as presented in the response optimization plot of Fig. **(6)**. In particular, the optimization utilizes the three base variables in the model, optimizing their combination. For this purpose, the individual desirability (d), referred to each single response factor, is first calculated; then, the composite desirability (D) is determined, combining the maximization and optimization points of the factor levels. The average response maximum (y) is described by dashed horizontal lines, while the optimum levels for the factors are represented by the vertical lines crossing through individual points.

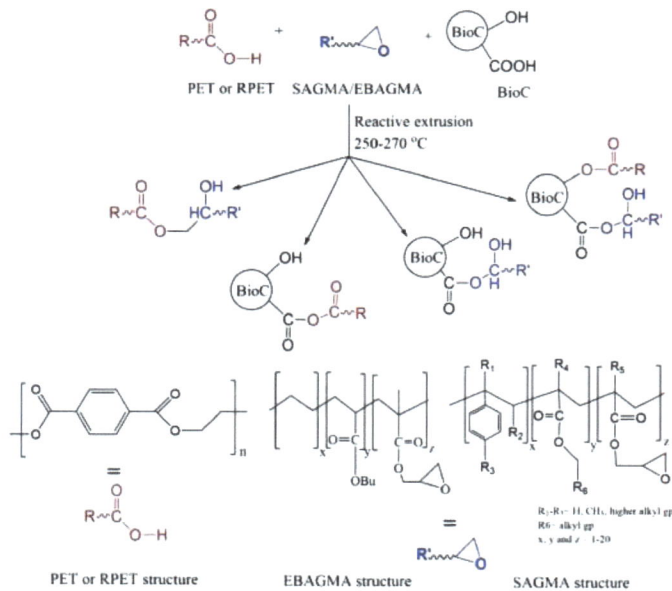

Fig. (5). Possible reactions between PET/RPET, SAGMA/EBAGMA, and Biochar (BioC) biocomposites. Reprinted from [65] under CC BY 4.0 license.

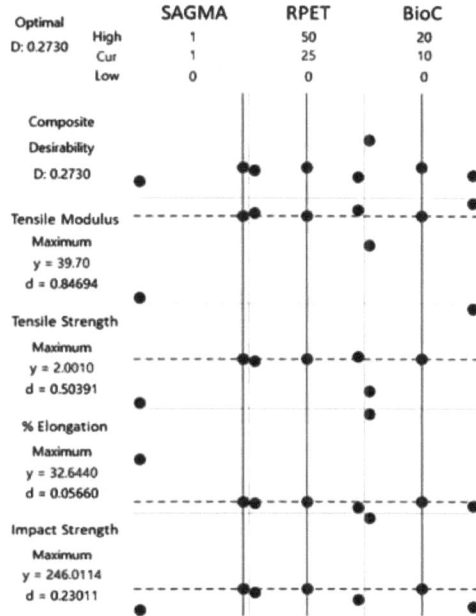

Fig. (6). Response optimization plot for the investigated composites. Reprinted from [65] under CC BY 4.0 license.

Quite recently, Chang and co-workers [66] exploited extrusion and subsequent injection molding for preparing **poly(butylene terephthalate)**-biochar composites. The biochar, obtained from miscanthus grass treated at 650°C, was incorporated into the polymer at various loadings, (5 - 25 wt.%). Aging tests performed at 155°C for 1000 hours allowed for assessing the durability of the mechanical features of the composites, which was compared with that of similar composites containing the same loading of either talc or glass fibers. Aging accounted for a rise in both tensile and flexural moduli of the investigated composites. Besides, the composites incorporating the biochar exhibited the highest lowering of mechanical strength after aging: this finding was ascribed to the lower adhesion, stiffness, and mechanical strength provided by the biochar, with respect to glass fibers or even talc.

Nagarajan and co-workers [67] investigated the effect of the size range (namely, 212–300, 150–212, 125–150, 75–125, 20–75, and <20 μm) of biochar obtained from low-temperature pyrolysis of Miscanthus on the mechanical performances of poly(trimethylene terephthalate)-poly(lactic acid) blends (70:30 wt. ratio), in the presence of ethylene methyl acrylate glycidyl methacrylate (15 wt.%) as a toughening agent (*i.e.*, chain extender). Both the biochar side and the chain extender were found to significantly affect the morphology of the obtained composites, as well as their mechanical behavior. More specifically, the most uniform dispersion was obtained for the systems embedding biochar particles with 20-75 µm size range. In addition, the presence of ethylene methyl acrylate glycidyl methacrylate was found to promote the formation of very fine morphologies, hence remarkably increasing the impact resistance of the composites that achieved 85 J/m (as assessed through Izod impact tests). Further, rheological tests revealed an increased complex viscosity with increasing the filler content, without affecting the shear-thinning behavior of the polymer matrix. Finally, increasing the processing temperature of the injection mold from 30 to 90°C accounted for an increased degree of crystallinity that enhanced both the flexural modulus and strength of the injection molded samples.

3.4. Biochar in other Thermoplastic Matrices

Polycarbonate is one of the engineering polymers that is gathering a lot of interest from the scientific community and is being exploited as an amorphous matrix for the preparation of biochar-based composites. Andrzejewsk and co-workers [68] exploited an extrusion/injection molding process for preparing polycarbonate-based composites, where the initial recycled carbon fiber content (20 wt.%) was progressively decreased and replaced with biochar (average particle size: about 400 µm) derived from the pyrolysis of Miscanthus at 900°C. The composites filled with 10 wt.% of both fillers exhibited remarkable enhancements in both tensile strength (+270%) and modulus (+35%) with respect to the counterpart containing 20 wt.% of biochar. Flexural tests revealed similar results, with around 305 and 16% increase in tensile strength and tensile modulus, respectively. The full replacement of recycled carbon fibers with biochar accounted for a further increase in tensile modulus; however, these composites showed a significant decrease in both ductility and impact strength.

Then, the same group succeeded in limiting the degradation phenomena that involve polycarbonate during melt processing [69]. To this aim, acrylonitrile butadiene styrene (ABS) was added to polycarbonate and mixed together with different fillers (*i.e.*, carbon fibers, biochar, and carbon fibers/biochar). Despite a worsening of the mechanical behavior and a lowering in the glass transition temperature as well, the presence of ABS was essential to prevent the hydrolytic degradation of the obtained materials.

Nan and co-workers [70] prepared **poly(vinyl alcohol)**-biochar composite films by solvent casting, at different filler loadings (namely, 2, 6, and 10 wt.%). The electrical conductivity showed an increasing trend with increasing the filler content, achieving $1.833 \cdot 10^{-6}$ S/cm in the presence of 10 wt.% of biochar. Further, the incorporation of the biochar accounted for increased Young's moduli and thermal stability but lowered tensile strength and both melting point and glass transition temperatures.

Pursuing this research, the same group [71] studied the suitability of poly(vinyl alcohol)-biochar composite films as pressure sensors. Increasing the biochar loading to 12 wt.% accounted for an increase in both electrical conductivity and piezoresistive effect, thanks to the increased conductive paths generated by the filler particles. In addition, a clear effect of increasing pressure from 0 to 358 kPa on the resistance of the composite films was observed: in fact, the resistance dropped down by 99% when 12 wt.% of filler was incorporated into the polymer matrix.

Recently, Diaz and co-workers [72] prepared blends of starch and polycaprolactone (50:50 wt.%) containing biochar derived from consumed coffee grounds at different loadings (ranging from 10 to 30 wt.%). 10 wt.% of the filler significantly affected the mechanical behavior, increasing the stiffness and slightly decreasing the tensile strength; conversely, further addition of biochar remarkably worsened the ductility of the composites. However, the authors succeeded in demonstrating the suitability of the designed system for thermoforming processes, showing its potential as a possible candidate material to replace traditional fossil-derived plastics in food service ware and packaging.

Among biodegradable materials other than starch, poly(3-hydroxybutyrate-c--3-hydroxyvalerate) (PHBV) was recently exploited for preparing composites containing biochar derived from Miscanthus (at different loadings, namely 10, 20, and 30 wt.%) [73]. As shown in Fig. (7), increasing the BC content accounted for an increase in the flexural modulus and a slight decrease in the flexural strength. Furthermore, the coefficient of linear thermal expansion significantly lowered with increasing the BC loading (Fig. 8), thus indicating an increased dimensional stability of the composites due to the incorporation of the stiff filler.

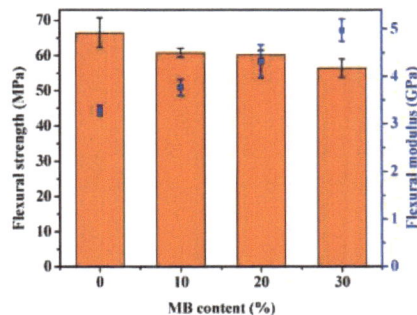

Fig. (7). Flexural strength and modulus of PHBV and its composites with biochar (MB). Reprinted from [73] under CC BY 4.0 license.

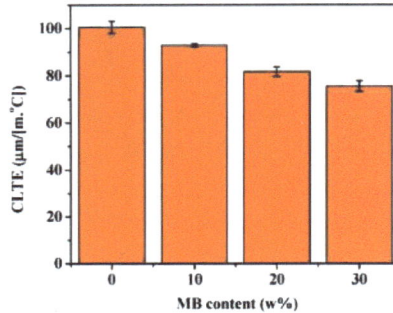

Fig. (8). Coefficient of linear thermal expansion (CLTE) of PHVH and its composites with biochar (MB). Reprinted from [73] under CC BY 4.0 license.

4. CARBONACEOUS FILLERS *VS.* BIOCHAR IN THE DESIGN OF POLYMER COMPOSITES

For several functional and structural applications, there exists a plethora of carbonaceous fillers ranging from nanosized objects (single-walled and multi-walled carbon nanotubes, graphenes) to carbon black, graphite, and carbon fibers. In this context, biochar can be considered a "new entry", quite recently discovered and suitable for applications that go beyond its consolidated use as a cheap carbon sequester and a natural adsorbent, as well as in soil remediation and amendment, or as a material suitable for electrochemical purposes [74 - 76]. Indeed, the pioneering work that appeared in 2015 [35] has paved the way for the exploitation of biochar for the design and preparation of novel polymer composites with functional or structural properties. The main advantages exhibited by this new filler can be summarized as follows:

- Biochar is bio-sourced and much cheaper with respect to conventional nanostructured carbon materials (like graphene and carbon nanotubes).

- It is compatible with several apolar to polar polymer matrices (both thermoplastic and thermosetting), notwithstanding the possibility of functionalizing it to make its dispersion easier.

- It well matches the current circular economy concept [77], as it is derived from by-products, biomasses, or wastes that are usually difficult to recover and valorize at the end of life.

- It allows for designing and producing composites with tunable properties that are strictly related to the structure, size, and size distribution, morphology, porosity, and possible functionalization of the filler.

However, biochar is currently suffering from different disadvantages that significantly limit its potential uses. First, biochar usually requires higher loadings in order to achieve performances comparable with nanostructured carbon materials. Further, despite all the current efforts, its yearly production is quite limited (and suitable for lab-scale activities) and poorly standardized: These factors are undoubtedly restricting the industrial utilization of biochar. Finally, biochar has to face competition with carbon black, the real competitor for the production of polymer composite materials. In fact, carbon black is very cheap, abundant, and with a robust industrial reputation that fully justifies its use for the production of commercial inks and tires, among others. On the other hand, unlike biochar, carbon black is fossil-derived, hence it possesses low sustainability and a not negligible environmental impact.

CONCLUSION AND FUTURE PERSPECTIVES

The present chapter has clearly demonstrated that biochar can be exploited as a multifunctional filler for the design of thermoplastic polymer composites showing tailored features. In particular, biochar can provide different polymer matrices with enhanced mechanical, electrical, thermal, and flame-retardant properties, among others. This way, it is possible to drastically lower the dependence on fossil-based fillers, valorizing, at the same time, post-production and post-industrial wastes and by-products at the end of life. Although the design and production of biochar-thermoplastic composites are currently at an early development stage, it is likely to be expected that this filler will acquire a key market space, at least partially replacing the conventional fillers already present and used. To this aim, it will be very important to further develop robust and reliable structure-property-production relationships for biochar, hence limiting the adverse effects exerted by the variability of the feedstock materials and better understanding of the effects of the experimental conditions adopted for the pyrolysis processes on the final carbonaceous products. In doing so, it will be possible to better foresee and control the overall features of the obtained polymer-biochar composites, hence widening their application fields.

ACKNOWLWDGEMENTS

The European Union—NextGenerationEU (National Sustainable Mobility Center CN00000023, Italian Ministry of University and Research Decree n. 1033-17/06/2022, Spoke 11—Innovative Materials & Lightweighting) is gratefully acknowledged. The opinions expressed are those of the author only and should not be considered as representative of the European Union or the European Commission's official position. Neither the European Union nor the European Commission can be held responsible for them.

REFERENCES

[1]   Yu, L.; Dean, K.; Li, L. Polymer blends and composites from renewable resources. *Prog. Polym. Sci.,* **2006**, *31*(6), 576-602.
[http://dx.doi.org/10.1016/j.progpolymsci.2006.03.002]

[2]   Iyer, K.A.; Zhang, L.; Torkelson, J.M. Direct use of natural antioxidant-rich agro-wastes as a thermal stabilizer for polymer: Processing and recycling. *ACS Sustain. Chem.& Eng.,* **2016**, *4*(3), 881-889.
[http://dx.doi.org/10.1021/acssuschemeng.5b00945]

[3]   Das, O.; Bhattacharyya, D.; Sarmah, A.K. Sustainable eco–composites obtained from waste derived biochar: A consideration in performance properties, production costs, and environmental impact. *J. Clean. Prod.,* **2016**, *129*, 159-168.
[http://dx.doi.org/10.1016/j.jclepro.2016.04.088]

[4]   Fiksel, J. Designing resilient, sustainable systems. *Environ. Sci. Technol.,* **2003**, *37*(23), 5330-5339.
[http://dx.doi.org/10.1021/es0344819] [PMID: 14700317]

[5]   Zimmerman, A.R. Abiotic and microbial oxidation of laboratory-produced black carbon (biochar). *Environ. Sci. Technol.,* **2010**, *44*(4), 1295-1301.
[http://dx.doi.org/10.1021/es903140c] [PMID: 20085259]

[6]   Wang, J.; Wang, S. Preparation, modification and environmental application of biochar: A review. *J. Clean. Prod.,* **2019**, *227*, 1002-1022.
[http://dx.doi.org/10.1016/j.jclepro.2019.04.282]

[7]   Cha, J.S.; Park, S.H.; Jung, S.C.; Ryu, C.; Jeon, J.K.; Shin, M.C.; Park, Y.K. Production and utilization of biochar: A review. *J. Ind. Eng. Chem.,* **2016**, *40*, 1-15.
[http://dx.doi.org/10.1016/j.jiec.2016.06.002]

[8]   Weber, K.; Quicker, P. Properties of biochar. *Fuel,* **2018**, *217*, 240-261.
[http://dx.doi.org/10.1016/j.fuel.2017.12.054]

[9]   Haris, M.; Hamid, Y.; Usman, M.; Wang, L.; Saleem, A.; Su, F.; Guo, J.; Li, Y. Crop-residues derived biochar: Synthesis, properties, characterization and application for the removal of trace elements in soils. *J. Hazard. Mater.,* **2021**, *416*, 126212.
[http://dx.doi.org/10.1016/j.jhazmat.2021.126212]

[10]   Sohi, S.P.; Krull, E.; Lopez-Capel, E.; Bol, R. A review of biochar and its use and function in soil. *Adv. Agron.,* **2010**, *105*, 47-82.
[http://dx.doi.org/10.1016/S0065-2113(10)05002-9]

[11]   Ferraro, G.; Pecori, G.; Rosi, L.; Bettucci, L.; Fratini, E.; Casini, D.; Rizzo, A.M.; Chiaramonti, D. Biochar from lab-scale pyrolysis: influence of feedstock and operational temperature. *Biomass Convers. Biorefin.,* **2021**, (Feb), 1-11.
[http://dx.doi.org/10.1007/s13399-021-01303-5]

[12]   Kameyama, K.; Miyamoto, T.; Iwata, Y. The preliminary study of water-retention related properties of biochar produced from various feedstock at different pyrolysis temperatures. *Materials (Basel),* **2019**, *12*(11), 1732.
[http://dx.doi.org/10.3390/ma12111732] [PMID: 31141965]

[13]   Dieguez-Alonso, A.; Funke, A.; Anca-Couce, A.; Rombolà, A.; Ojeda, G.; Bachmann, J.; Behrendt, F. Towards biochar and hydrochar engineering - Influence of process conditions on surface physical and chemical properties, thermal stability, nutrient availability, toxicity and wettability. *Energies,* **2018**, *11*(3), 496.
[http://dx.doi.org/10.3390/en11030496]

[14]   Cetin, E.; Moghtaderi, B.; Gupta, R.; Wall, T.F. Influence of pyrolysis conditions on the structure and gasification reactivity of biomass chars. *Fuel,* **2004**, *83*(16), 2139-2150.
[http://dx.doi.org/10.1016/j.fuel.2004.05.008]

[15]   Zhao, B.; O'Connor, D.; Zhang, J.; Peng, T.; Shen, Z.; Tsang, D.C.W.; Hou, D. Effect of pyrolysis temperature, heating rate, and residence time on rapeseed stem derived biochar. *J. Clean. Prod.,* **2018,** *174,* 977-987.
[http://dx.doi.org/10.1016/j.jclepro.2017.11.013]

[16]   Angın, D.; Altintig, E.; Köse, T.E. Influence of process parameters on the surface and chemical properties of activated carbon obtained from biochar by chemical activation. *Bioresour. Technol.,* **2013,** *148,* 542-549.
[http://dx.doi.org/10.1016/j.biortech.2013.08.164] [PMID: 24080293]

[17]   Mujtaba Munir, M.A.; Yousaf, B.; Ali, M.U.; Dan, C.; Abbas, Q.; Arif, M.; Yang, X. *In situ* synthesis of micro-plastics embedded sewage-sludge co-pyrolyzed biochar: Implications for the remediation of Cr and Pb availability and enzymatic activities from the contaminated soil. *J. Clean. Prod.,* **2021,** *302,* 127005.
[http://dx.doi.org/10.1016/j.jclepro.2021.127005]

[18]   El-Naggar, A.; Lee, S.S.; Rinklebe, J.; Farooq, M.; Song, H.; Sarmah, A.K.; Zimmerman, A.R.; Ahmad, M.; Shaheen, S.M.; Ok, Y.S. Biochar application to low fertility soils: A review of current status, and future prospects. *Geoderma,* **2019,** *337,* 536-554.
[http://dx.doi.org/10.1016/j.geoderma.2018.09.034]

[19]   Li, X.; Zhang, J.; Liu, B.; Su, Z. A critical review on the application and recent developments of post-modified biochar in supercapacitors. *J. Clean. Prod.,* **2021,** *310,* 127428.
[http://dx.doi.org/10.1016/j.jclepro.2021.127428]

[20]   Mehdi, R.; Khoja, A.H.; Naqvi, S.R.; Gao, N.; Amin, N.A.S. A Review on Production and Surface Modifications of Biochar Materials *via* Biomass Pyrolysis Process for Supercapacitor Applications. *Catalysts,* **2022,** *12*(7), 798.
[http://dx.doi.org/10.3390/catal12070798]

[21]   Lee, J.; Kim, K.H.; Kwon, E.E. Biochar as a Catalyst. *Renew. Sustain. Energy Rev.,* **2017,** *77,* 70-79.
[http://dx.doi.org/10.1016/j.rser.2017.04.002]

[22]   Md Khudzari, J.; Gariépy, Y.; Kurian, J.; Tartakovsky, B.; Raghavan, G.S.V. Effects of biochar anodes in rice plant microbial fuel cells on the production of bioelectricity, biomass, and methane. *Biochem. Eng. J.,* **2019,** *141,* 190-199.
[http://dx.doi.org/10.1016/j.bej.2018.10.012]

[23]   Das, C.; Tamrakar, S.; Kiziltas, A.; Xie, X. Incorporation of Biochar to Improve Mechanical, Thermal and Electrical Properties of Polymer Composites. *Polymers (Basel),* **2021,** *13*(16), 2663.
[http://dx.doi.org/10.3390/polym13162663] [PMID: 34451201]

[24]   Giorcelli, M.; Bartoli, M. Development of Coffee Biochar Filler for the Production of Electrical Conductive Reinforced Plastic. *Polymers (Basel),* **2019,** *11*(12), 1916.
[http://dx.doi.org/10.3390/polym11121916] [PMID: 31766390]

[25]   Giorcelli, M.; Khan, A.; Pugno, N.M.; Rosso, C.; Tagliaferro, A. Biochar as a cheap and environmental friendly filler able to improve polymer mechanical properties. *Biomass Bioenergy,* **2019,** *120,* 219-223.
[http://dx.doi.org/10.1016/j.biombioe.2018.11.036]

[26]   Purakayastha, T.J.; Bera, T.; Bhaduri, D.; Sarkar, B.; Mandal, S.; Wade, P.; Kumari, S.; Biswas, S.; Menon, M.; Pathak, H.; Tsang, D.C.W. A review on biochar modulated soil condition improvements and nutrient dynamics concerning crop yields: Pathways to climate change mitigation and global food security. *Chemosphere,* **2019,** *227,* 345-365.
[http://dx.doi.org/10.1016/j.chemosphere.2019.03.170] [PMID: 30999175]

[27]   Soni, B.; Karmee, S.K. Towards a continuous pilot scale pyrolysis based biorefinery for production of biooil and biochar from sawdust. *Fuel,* **2020,** *271,* 117570.
[http://dx.doi.org/10.1016/j.fuel.2020.117570]

[28]   Li, F.; Chen, J.; Hu, X.; He, F.; Bean, E.; Tsang, D.C.W.; Ok, Y.S.; Gao, B. Applications of carbonaceous adsorbents in the remediation of polycyclic aromatic hydrocarbon-contaminated sediments: A review. *J. Clean. Prod.,* **2020**, *255*, 120263.
        [http://dx.doi.org/10.1016/j.jclepro.2020.120263]

[29]   Leng, L.; Xiong, Q.; Yang, L.; Li, H.; Zhou, Y.; Zhang, W.; Jiang, S.; Li, H.; Huang, H. An overview on engineering the surface area and porosity of biochar. *Sci. Total Environ.,* **2021**, *763*, 144204.
        [http://dx.doi.org/10.1016/j.scitotenv.2020.144204] [PMID: 33385838]

[30]   Wang, L.; Ok, Y.S.; Tsang, D.C.W.; Alessi, D.S.; Rinklebe, J.; Wang, H.; Mašek, O.; Hou, R.; O'Connor, D.; Hou, D. New trends in biochar pyrolysis and modification strategies: feedstock, pyrolysis conditions, sustainability concerns and implications for soil amendment. *Soil Use Manage.,* **2020**, *36*(3), 358-386.
        [http://dx.doi.org/10.1111/sum.12592]

[31]   Zhang, Z.; Zhu, Z.; Shen, B.; Liu, L. Insights into biochar and hydrochar production and applications: A review. *Energy,* **2019**, *171*, 581-598.
        [http://dx.doi.org/10.1016/j.energy.2019.01.035]

[32]   Bolan, N.; Hoang, S.A.; Beiyuan, J.; Gupta, S.; Hou, D.; Karakoti, A.; Joseph, S.; Jung, S.; Kim, K.H.; Kirkham, M.B.; Kua, H.W.; Kumar, M.; Kwon, E.E.; Ok, Y.S.; Perera, V.; Rinklebe, J.; Shaheen, S.M.; Sarkar, B.; Sarmah, A.K.; Singh, B.P.; Singh, G.; Tsang, D.C.W.; Vikrant, K.; Vithanage, M.; Vinu, A.; Wang, H.; Wijesekara, H.; Yan, Y.; Younis, S.A.; Van Zwieten, L. Multifunctional applications of biochar beyond carbon storage. *Int. Mater. Rev.,* **2022**, *67*(2), 150-200.
        [http://dx.doi.org/10.1080/09506608.2021.1922047]

[33]   Omoriyekomwan, J.E.; Tahmasebi, A.; Zhang, J.; Yu, J. Formation of hollow carbon nanofibers on bio-char during microwave pyrolysis of palm kernel shell. *Energy Convers. Manage.,* **2017**, *148*, 583-592.
        [http://dx.doi.org/10.1016/j.enconman.2017.06.022]

[34]   Lee, X.J.; Ong, H.C.; Gan, Y.Y.; Chen, W.H.; Mahlia, T.M.I. State of art review on conventional and advanced pyrolysis of macroalgae and microalgae for biochar, bio-oil and bio-syngas production. *Energy Convers. Manage.,* **2020**, *210*, 112707.
        [http://dx.doi.org/10.1016/j.enconman.2020.112707]

[35]   Das, O.; Sarmah, A.K.; Bhattacharyya, D. A novel approach in organic waste utilization through biochar addition in wood/polypropylene composites. *Waste Manag.,* **2015**, *38*, 132-140.
        [http://dx.doi.org/10.1016/j.wasman.2015.01.015] [PMID: 25677179]

[36]   Das, O.; Sarmah, A.K.; Bhattacharyya, D. Biocomposites from waste derived biochars: Mechanical, thermal, chemical, and morphological properties. *Waste Manag.,* **2016**, *49*, 560-570.
        [http://dx.doi.org/10.1016/j.wasman.2015.12.007] [PMID: 26724232]

[37]   Das, O.; Bhattacharyya, D.; Hui, D.; Lau, K-T. "Mechanical and flammability characterisations of biochar/polypropylene biocomposites", *Compos. B Eng.,* **2016**, *106*, 120-128.

[38]   Gezahegn, S.; Lai, R.; Huang, L.; Chen, L.; Huang, F.; Blozowski, N.; Thomas, S.C.; Sain, M.; Tjong, J.; Jaffer, S.; Behravesh, A.; Weimin, Y. Porous graphitic biocarbon and reclaimed carbon fiber derived environmentally benign lightweight composites. *Sci. Total Environ.,* **2019**, *664*, 363-373.
        [http://dx.doi.org/10.1016/j.scitotenv.2019.01.408] [PMID: 30743128]

[39]   Das, O.; Kim, N.K.; Sarmah, A.K.; Bhattacharyya, D. Development of waste based biochar/wool hybrid biocomposites: Flammability characteristics and mechanical properties. *J. Clean. Prod.,* **2017**, *144*, 79-89.
        [http://dx.doi.org/10.1016/j.jclepro.2016.12.155]

[40]   Das, O.; Kim, N.K.; Kalamkarov, A.L.; Sarmah, A.K.; Bhattacharyya, D. Biochar to the rescue: Balancing the fire performance and mechanical properties of polypropylene composites. *Polym. Degrad. Stabil.,* **2017**, *144*, 485-496.
        [http://dx.doi.org/10.1016/j.polymdegradstab.2017.09.006]

[41]   Paleri, D.M.; Rodriguez-Uribe, A.; Misra, M.; Mohanty, A.K. Pyrolyzed biomass from corn ethanol industry coproduct and their polypropylene-based composites: effect of heat treatment temperature on performance of the biocomposites. *Compos., Part B Eng.,* **2021**, *215*, 108714.
[http://dx.doi.org/10.1016/j.compositesb.2021.108714]

[42]   Ayadi, R.; Koubaa, A.; Braghiroli, F.; Migneault, S.; Wang, H.; Bradai, C. Effect of the Pyro-Gasification Temperature of Wood on the Physical and Mechanical Properties of Biochar-Polymer Biocomposites. *Materials (Basel),* **2020**, *13*(6), 1327.
[http://dx.doi.org/10.3390/ma13061327] [PMID: 32183329]

[43]   Alghyamah, A.A.; Yagoub Elnour, A.; Shaikh, H.; Haider, S.; Manjaly Poulose, A.; Al-Zahrani, S.M.; Almasry, W.A.; Young Park, S. Biochar/polypropylene composites: A study on the effect of pyrolysis temperature on crystallization kinetics, crystalline structure, and thermal stability. *J. King Saud Univ. Sci.,* **2021**, *33*(4), 101409.
[http://dx.doi.org/10.1016/j.jksus.2021.101409]

[44]   Wang, X.; Sotoudehniakarani, F.; Yu, Z.; Morrell, J.J.; Cappellazzi, J.; McDonald, A.G. Evaluation of corrugated cardboard biochar as reinforcing fiber on properties, biodegradability and weatherability of wood-plastic composites. *Polym. Degrad. Stabil.,* **2019**, *168*, 108955.
[http://dx.doi.org/10.1016/j.polymdegradstab.2019.108955]

[45]   Li, S.; Huang, A.; Chen, Y-J.; Li, D.; Turng, L-S. "Highly filled biochar/ultra-high molecular weight polyethylene/linear low density polyethylene composites for high-performance electromagnetic interference shielding". *Compos. B Eng.,* **2018**, *153*, 277-284.

[46]   Li, S.; Li, X.; Deng, Q.; Li, D. Three kinds of charcoal powder reinforced ultra-high molecular weight polyethylene composites with excellent mechanical and electrical properties. *Mater. Des.,* **2015**, *85*, 54-59.
[http://dx.doi.org/10.1016/j.matdes.2015.06.163]

[47]   Li, S.; Li, X.; Chen, C.; Wang, H.; Deng, Q.; Gong, M.; Li, D. Development of electrically conductive nano bamboo charcoal/ultra-high molecular weight polyethylene composites with a segregated network. *Compos. Sci. Technol.,* **2016**, *132*, 31-37.
[http://dx.doi.org/10.1016/j.compscitech.2016.06.010]

[48]   Li, S.; Xu, Y.; Jing, X.; Yilmaz, G.; Li, D.; Turng, L-S. "Effect of carbonization temperature on mechanical properties and biocompatibility of biochar/ultra-high molecular weight polyethylene composites", *Compos. B Eng.,* **2020**, *196*, 108120.

[49]   Zhang, Q.; Xu, H.; Lu, W.; Zhang, D.; Ren, X.; Yu, W.; Wu, J.; Zhou, L.; Han, X.; Yi, W.; Lei, H. Properties evaluation of biochar/high-density polyethylene composites: Emphasizing the porous structure of biochar by activation. *Sci. Total Environ.,* **2020**, *737*, 139770.
[http://dx.doi.org/10.1016/j.scitotenv.2020.139770] [PMID: 32512307]

[50]   Zhang, Q.; Khan, M.U.; Lin, X.; Cai, H.; Lei, H. "Temperature varied biochar as a reinforcing filler for high-density polyethylene composites", *Compos. B Eng.,* **2019**, *175*, 107151.

[51]   Arrigo, R.; Jagdale, P.; Bartoli, M.; Tagliaferro, A.; Malucelli, G. Structure–Property Relationships in Polyethylene-Based Composites Filled with Biochar Derived from Waste Coffee Grounds. *Polymers (Basel),* **2019**, *11*(8), 1336.
[http://dx.doi.org/10.3390/polym11081336] [PMID: 31409023]

[52]   Ogunsona, E.O.; Misra, M.; Mohanty, A.K. Sustainable biocomposites from biobased polyamide 6,10 and biocarbon from pyrolyzed miscanthus fibers. *J. Appl. Polym. Sci.,* **2017**, *134*(4), 44221.
[http://dx.doi.org/10.1002/app.44221]

[53]   Ogunsona, E.O.; Misra, M.; Mohanty, A.K. "Impact of interfacial adhesion on the microstructure and property variations of biocarbons reinforced nylon 6 biocomposites". *Compos. -. A: Appl. Sci. Manuf.,* **2017**, *98*, 32-44.

[54]   Watt, E.; Abdelwahab, M.A.; Mohanty, A.K.; Misra, M. "Biocomposites from biobased polyamide

4,10 and waste corn cob based biocarbon", *Compos. -. A: Appl. Sci. Manuf.,* **2021**, *145*, 106340.

[55]   Murariu, M.; Dubois, P. PLA composites: From production to properties. *Adv. Drug Deliv. Rev.,* **2016**, *107*, 17-46.
[http://dx.doi.org/10.1016/j.addr.2016.04.003] [PMID: 27085468]

[56]   Ho, M.P.; Lau, K.T.; Wang, H.; Hui, D. "Improvement on the properties of polylactic acid (PLA) using bamboo charcoal particles", *Compos. B Eng.,* **2015**, *81*, 14-25.

[57]   Qian, S.; Sheng, K.; Yao, W.; Yu, H. Poly(lactic acid) biocomposites reinforced with ultrafine bamboo-char: Morphology, mechanical, thermal, and water absorption properties. *J. Appl. Polym. Sci.,* **2016**, *133*(20), n/a.
[http://dx.doi.org/10.1002/app.43425]

[58]   Salak, F.; Uemura, S.; Sugimoto, K. Thermal pretreatment of kudzu biomass (*pueraria lobata*) as filler in cost-effective pla biocomposite fabrication process. *Polym. Eng. Sci.,* **2015**, *55*(2), 340-348.
[http://dx.doi.org/10.1002/pen.23909]

[59]   Qian, S.; Yan, W.; Zhu, S.; Fontanillo Lopez, C.A.; Sheng, K. Surface modification of bamboo-char and its reinforcement in PLA biocomposites. *Polym. Compos.,* **2018**, *39*, E633-E639.
[http://dx.doi.org/10.1002/pc.24800]

[60]   Nizamuddin, S.; Jadhav, A.; Qureshi, S.S.; Baloch, H.A.; Siddiqui, M.T.H.; Mubarak, N.M.; Griffin, G.; Madapusi, S.; Tanksale, A.; Ahamed, M.I. Synthesis and characterization of polylactide/rice husk hydrochar composite. *Sci. Rep.,* **2019**, *9*(1), 5445.
[http://dx.doi.org/10.1038/s41598-019-41960-1] [PMID: 30931991]

[61]   Arrigo, R.; Bartoli, M.; Malucelli, G. Poly(lactic Acid)–Biochar Biocomposites: Effect of Processing and Filler Content on Rheological, Thermal, and Mechanical Properties. *Polymers (Basel),* **2020**, *12*(4), 892.
[http://dx.doi.org/10.3390/polym12040892] [PMID: 32290601]

[62]   Oliveira, M.; Santos, E.; Araújo, A.; Fechine, G.J.M.; Machado, A.V.; Botelho, G. The role of shear and stabilizer on PLA degradation. *Polym. Test.,* **2016**, *51*, 109-116.
[http://dx.doi.org/10.1016/j.polymertesting.2016.03.005]

[63]   Sheng, K.; Zhang, S.; Qian, S.; Lopez, C.A.F. "High-toughness PLA/Bamboo cellulose nanowhiskers bionanocomposite strengthened with silylated ultrafine bamboo-char", *Compos. B Eng.,* **2019**, *165*, 174-182.

[64]   Tolvanen, J.; Hannu, J.; Hietala, M.; Kordas, K.; Jantunen, H. Biodegradable multiphase poly(lactic acid)/biochar/graphite composites for electromagnetic interference shielding. *Compos. Sci. Technol.,* **2019**, *181*, 107704.
[http://dx.doi.org/10.1016/j.compscitech.2019.107704]

[65]   Snowdon, M.R.; Abdelwahab, M.; Mohanty, A.K.; Misra, M. Mechanical optimization of virgin and recycled poly(ethylene terephthalate) biocomposites with sustainable biocarbon through a factorial design. *Results in Materials,* **2020**, *5*, 100060.
[http://dx.doi.org/10.1016/j.rinma.2020.100060]

[66]   Chang, B.P.; Mohanty, A.K.; Misra, M. Sustainable biocarbon as an alternative of traditional fillers for poly(butylene terephthalate)-based composites: Thermo-oxidative aging and durability. *J. Appl. Polym. Sci.,* **2019**, *136*(27), 47722.
[http://dx.doi.org/10.1002/app.47722]

[67]   Nagarajan, V.; Mohanty, A.K.; Misra, M. Biocomposites with Size-Fractionated Biocarbon: Influence of the Microstructure on Macroscopic Properties. *ACS Omega,* **2016**, *1*(4), 636-647.
[http://dx.doi.org/10.1021/acsomega.6b00175] [PMID: 31457153]

[68]   Andrzejewski, J.; Misra, M.; Mohanty, A.K. Polycarbonate biocomposites reinforced with a hybrid filler system of recycled carbon fiber and biocarbon: Preparation and thermomechanical characterization. *J. Appl. Polym. Sci.,* **2018**, *135*(28), 46449.

[http://dx.doi.org/10.1002/app.46449]

[69]   Andrzejewski, J.; Mohanty, A.K.; Misra, M. Development of hybrid composites reinforced with biocarbon/carbon fiber system. The comparative study for PC, ABS and PC/ABS based materials. *Compos., Part B Eng.,* **2020**, *200*, 108319.
[http://dx.doi.org/10.1016/j.compositesb.2020.108319]

[70]   Nan, N.; DeVallance, D.B.; Xie, X.; Wang, J. The effect of bio-carbon addition on the electrical, mechanical, and thermal properties of polyvinyl alcohol/biochar composites. *J. Compos. Mater.,* **2016**, *50*(9), 1161-1168.
[http://dx.doi.org/10.1177/0021998315589770]

[71]   Nan, N.; DeVallance, D.B. Development of poly(vinyl alcohol)/wood-derived biochar composites for use in pressure sensor applications. *J. Mater. Sci.,* **2017**, *52*(13), 8247-8257.
[http://dx.doi.org/10.1007/s10853-017-1040-7]

[72]   Diaz, C.A.; Shah, R.K.; Evans, T.; Trabold, T.A.; Draper, K. Thermoformed Containers Based on Starch and Starch/Coffee Waste Biochar Composites. *Energies,* **2020**, *13*(22), 6034.
[http://dx.doi.org/10.3390/en13226034]

[73]   Li, Z.; Reimer, C.; Wang, T.; Mohanty, A.K.; Misra, M. Thermal and Mechanical Properties of the Biocomposites of Miscanthus Biocarbon and Poly(3-Hydroxybutyrate-co-3-Hydroxyvalerate) (PHBV). *Polymers (Basel),* **2020**, *12*(6), 1300.
[http://dx.doi.org/10.3390/polym12061300] [PMID: 32517200]

[74]   Ghaedi, M.; Mazaheri, H.; Khodadoust, S.; Hajati, S.; Purkait, M.K. Application of central composite design for simultaneous removal of methylene blue and Pb^{2+} ions by walnut wood activated carbon. *Spectrochim. Acta A Mol. Biomol. Spectrosc.,* **2015**, *135*, 479-490.
[http://dx.doi.org/10.1016/j.saa.2014.06.138] [PMID: 25113736]

[75]   Ghaedi, M.; Nasab, A.G.; Khodadoust, S.; Rajabi, M.; Azizian, S. Application of activated carbon as adsorbents for efficient removal of methylene blue: Kinetics and equilibrium study. *J. Ind. Eng. Chem.,* **2014**, *20*(4), 2317-2324.
[http://dx.doi.org/10.1016/j.jiec.2013.10.007]

[76]   Lee, H.W.; Kim, Y.M.; Kim, S.; Ryu, C.; Park, S.H.; Park, Y.K. Review of the use of activated biochar for energy and environmental applications. *Carbon Lett.,* **2018**, *26*, 1-10.

[77]   Mhatre, P.; Panchal, R.; Singh, A.; Bibyan, S. A systematic literature review on the circular economy initiatives in the European Union. *Sustainable Production and Consumption,* **2021**, *26*, 187-202.
[http://dx.doi.org/10.1016/j.spc.2020.09.008]

CHAPTER 3

Animal-Based Biochar Reinforced Polymer Composites

Radhika Mandala[1,2,*], **B. Anjaneya Prasad**[1] and **Suresh Akella**[3]

[1] *Department of Mechanical Engineering, Jawaharlal Nehru Technological University, Hyderabad, Telangana 500085, India*

[2] *Department of Mechanical Engineering, Vignan Institute of Technology and Science, Deshmukhi, Hyderabad, Telangana 508284, India*

[3] *Department of Mechanical Engineering Sreyas Institute of Engineering and Technology, Naderagul, Hyderabad, Telangana 500068, India*

Abstract: Biomass-derived waste management has become an increasingly pressing concern due to rising levels of environmental issues. As a result, interest has risen in finding ways to turn biomass wastes into useful products. The conversion of biowaste into biochar is one of the efficient and environmentally friendly methods of disposing of biowaste. Developing polymer composites by reinforcing biochar as the filler material is gaining popularity due to their affordability and exceptional thermal and mechanical properties. Animal waste is one of the biomass wastes that can be converted into biochar and can be used in various applications. This review work aimed at synthesizing biochar from animal wastes, preparing polymer composites, and analyzing the thermo-mechanical properties. This review also focuses on various animal feedstocks for the synthesis of biocarbon and methods to fabricate polymer composites. The biocarbon-induced polymer composites showed an improvement in mechanical and thermal properties with varying percentages of loading.

Keywords: Animal Waste, Biochar, Polymer Nanocomposite, Pyrolysis.

1. INTRODUCTION

With the world's population expanding, the demand for goods, energy, and food is always increasing. Humans need a lot of resources for their survival, producing millions of tons of waste, which has a negative influence on our ecosystem [1]. It is predicted that by 2050, worldwide production of municipal solid garbage will increase to 3.4 billion metric tonnes, representing an increase of almost 70%. There are no signs that the tremendous rise in waste production that has occurred

---

* **Corresponding author Radhika Mandala:** Department of Mechanical Engineering, Jawaharlal Nehru Technological University, Hyderabad, Telangana, 500085, India; E-mail: mandalaradhika77@gmail.com

Deepa Kodali & Vijaya Rangari (Eds.)

in recent decades will abate. Dealing with this expanding volume of biowaste poses one of the greatest challenges to the sustainability of human civilization. Consequently, it is essential that biowaste should be disposed of efficiently and sustainably [2]. The conversion of biowaste into biocarbon is one of the efficient and environmentally friendly methods of disposing of biowaste that fascinates researchers and enterprises. Biocarbon, also known as biochar is a porous solid, with high carbon content, formed by biomass thermal decomposition in a reactor with little or no availability of air. In view of the rapid depletion of non-renewable resources and the increase of greenhouse effects, biochar became an undeniable resource to substitute fossil fuels and thus tailor the needs by serving as a filler in the polymer matrix. Owing to its affordability, extensive availability, and eco-friendliness, biocarbon prepared from inexpensive, abundant, and eco-friendly resources is gaining popularity in a wide variety of emerging industries including agriculture, environment, energy storage, biomedical, drug delivery, structural, materials and filtration [3]. Any affordable substance that contains a high carbon content and minimal inorganic matter can be utilized as a feedstock for producing biochar [4]. The biocarbon synthesized from biowaste makes it viable for the reinforcement of bio-based fillers with synthetic as well as natural polymers.

The polymer composite consists of a polymer as the matrix phase and the fibre or particles as the reinforcement phase. The particles can be macro/micro/nano size. Polymer composites are popular due to their low density, high specific strength, high specific stiffness, effortless fabrication, and cost-effectiveness. Reinforced polymers, which offer better properties than pure polymers, have attracted a lot more attention in recent years. A polymer nanocomposite is a composite material consisting of polymers of thermoplastics, thermosets, or elastomers reinforced with nanoparticles of sizes less than 100nm diameter with aspect ratios greater than 300 [5]. It is a multiphase solid material in which one of the phases has one, two or three-dimensions nanofillers. Nanoparticles have a remarkably high surface-to-volume ratio which significantly alters the inherent properties of a polymer when compared with their bulk-sized reinforcement [6]. Nowadays, nano-sized carbonaceous fillers including carbon nanotubes (CNTs), graphene, carbon nanospheres (CNSs), *etc.* are being used to fortify a wide range of polymer composites, with the goal of enhancing their mechanical and functional qualities [7]. However, these carbon nano fillers' extensive application has been severely hampered by their high manufacturing costs, limited production volumes, the need for modern instruments for synthesis, and environmentally hazardous preparation techniques [8]. As a result, there is a growing demand for nanostructured carbons made from inexpensive, plentiful, and environmentally friendly materials including vegetable biomass and agricultural waste, animal waste, and by-products. Nano-structured carbons are synthesized by thermal decomposition of biowaste. A lot of research is going on polymer nanocomposites reinforcing biocarbon derived from vegetable biomass and agriculture waste. Relatively limited research has been done on polymer nanocomposite reinforcing biocarbon derived from animal waste. Hence, this review focuses on the polymer composites reinforced with animal-based biochar of micro/macro/nano size.

2. OVERVIEW OF CARBON

Carbon is one of the exciting elements, with the ability to form a vast array of configurations that often exhibit a wide variety of properties [9]. The varieties of configurations are possible due to the covalent bonds with other carbon atoms and also with metallic or non-metallic elements in different hybridization states (sp, sp2, or sp3). Carbon materials are categorized based on carbon-carbon bonds of the hybrid orbitals of sp3, planar sp2 +π, curved sp2 +π p, and sp +2π [10]. Diamond, graphite, fullerenes, and carbynes are allotropes of carbon that primarily include either of these hybrid orbitals. Diamond is important for its carbon-containing sp3 hybridization. Graphite with its sp2 hybridization and layered graphene sheets has been identified as having the softest interior nature of all known materials. Also, graphite is an excellent conductor of heat, an opaque material, a good lubricant, and a good conductor of energy. In addition to diamond and graphite, other carbon allotropes include Buck balls (fullerenes), amorphous carbon, glassy carbon, carbon nanofoam, and carbon nanotubes. Carbonaceous materials are favoured and are one of the most sought-after materials due to their versatile nanostructure, outstanding mechanical, electrochemical, thermal, chemical stability, and electrical characteristics. Due to its diverse dimensionality (0 to 3D), it is a functional material that meets a variety of requirements for various applications. It can be obtained in various forms including powders, fibres, composites, tubes, felts, mats, monoliths, and foils [11]. Because of its exceptional characteristics, it is an appealing candidate for use in a wide range of applications, some of which are found in the fields of electronics, electrochemistry and sensing, energy, biomedicine, aerospace, automobiles, and the environment [12]. The high cost of manufacture and the material's reliance on fossil fuels pose the greatest barrier to the widespread use of carbon-based materials. Due to these reasons, researchers have recently become interested in eco-friendly carbon materials made from renewable resources that can be mass-produced. Biocarbon, also known as biochar, has gained widespread attention as a new sustainable material for a variety of commercial applications [13]. Biocarbon made from renewable resources is becoming increasingly popular because it avoids the environmental consequences associated with unreplenishable sources of carbon, like fossil fuels, coal, oil, and gas, in addition to being extremely cost-effective [1].

2.1. Sources of Biocarbon

An organic waste or waste material that decomposes under aerobic (oxygen-containing) or anaerobic (oxygen-deficient) conditions is named as biowaste. These waste materials include agriculture, animal waste, food waste, kitchen waste, sewage sludge, agro-industries, forestry residues, and other organic industrial wastes [14]. It is important to manage biowaste properly because it contributes to olfactive annoyance which attracts pests and other disease-carrying animals like rodents as well as generates toxic leachate that can contaminate surface and groundwater. In addition, uncontrolled biowaste disposal contributes to global warming by emitting methane [15]. Therefore, the waste obtained from biomass has emerged as a budding source of raw materials for the manufacture of biocarbon because of its affordability, ease of availability and ultimately capable of turning trash into an income. Animal waste, in particular, is a desirable biomass feedstock since it contains a wealth of nutrients and may be used to generate renewable energy.

The fishery, meat, leather, and poultry sectors are the most responsible industries for producing animal waste. These industries generate wastes such as fish scales, fish bones, crab shells, prawn shells, bovine manure, tannery scraps, chicken feathers, and other animal by-products. The increased demand from consumers has caused the poultry and meat sectors to rapidly increase their output of inedible by-products such as bones, feathers, tendons, and skins. The typical treatments of animal waste are burning them or composting them with manure, although both of these methods have drawbacks, including significant energy consumption, carbon dioxide emissions, an unpleasant smell from the H_2S produced during composting, and others. In an effort to combat the issues described above, recent research has centered on creating sustainable significant applications of animal waste, like the fabrication of carbon fibre from collagen [16], biogas [17], and biodiesel [18]. There is more focus on the production of biochar [19 - 21] from animal waste. The various sources of animal waste for biochar include chicken feathers, eggshells, chicken litter, fish scales, shrimp shells, crab shells, and cattle bones. The biocarbon produced can be used for various applications which include electrode materials for supercapacitors and Li-ion batteries, as an adsorber, for making polymer composites, *etc.*, which is shown in Table **1**.

Table 1. Various animal feed stocks, methods to produce biochar and its applications.

Feedstock	Biocarbon Synthesis Technique	Temperature in °C	Furnace Used	Application	Refs.
Chicken feathers	Pyrolysis	300&600	Tubular furnace	Polymer composite	[24]
		500	Continuous Pyrolyzer	Polymer composite	[25]
		1100	Ni-based superalloy high-pressure/temperature reactor	3Dprinted electrode	[26]
Shrimp shells	Pyrolysis	500	Continuous Pyrolyzer	Polymer composite	[25]
Chicken litter	Pyrolysis	450	Auger Reactor	Polymer composite	[27]
Eggshell	Carbonization	850	Muffle furnace	Polymer composite (Thermoset polymer)	[29, 30]
Crab shell	Carbonization	500	Tube furnace	Electrode Material for Supercapacitor	[33]
Chicken bones	pyrolysis	800	Bench Reactor	fuchsine adsorption	[34]
Cattle bones	Carbonization	600-1200	-	Electrode materials for Li-ion batteries and supercapacitor	[18]
Prawn shells	Hydrothermal carbonization	180	Muffle furnace	Carbon dots reinforced in epoxy polymer coatings	[31]
Ox horn	Carbonization	700	Horizontal tube furnace	Activated carbon	[35]
Animal bones +Rice straw	Pyrolysis	550	Semi-automatic pyrolyzer	Biochar for plant growth	[28]

2.2. Synthesis of Animal-based Biocarbon

Biocarbon can be synthesized from biomass or biomass waste by various processes named pyrolysis, carbonization, hydrothermal carbonization, gasification, and torrefaction. The most common technology used is pyrolysis. Pyrolysis is the high-temperature thermochemical disintegration of organic material in an inert or oxygen-free atmosphere [22]. Traditionally, this approach has been used to separate petroleum products, but now this approach is gaining popularity as an economically and environmentally beneficial way of transforming biowaste into biocarbon. During the pyrolysis process, moisture and volatile substances are driven off, leaving behind a carbon-rich solid residue known as biochar. Different types of reactors, including fixed bed, fluidized bed, tubular, and Auger reactors, can be used to conduct the pyrolysis process [23]. Li *et al.* [24] synthesized biocarbon by pyrolysis technique from chicken feathers (ChF) and chicken feathers (ChF) powder. The ChF and ChF powder were pyrolyzed at 300°C and 600°C in a tubular furnace with a holding period of 30min. In another study, Rodriguez *et al.* [25] synthesized biocarbon from chicken feathers and shrimp shells by pyrolysis process using a continuous pyrolyzer at 500°C. Mohammed *et al.* [26] synthesized biocarbon by pyrolyzing chicken feather powder at a temperature of 1100 °C through a ramping rate of 10°C /min using an autogenic pressure reactor. The process diagram of the synthesis of carbon is shown in Fig. (**1**). The chicken litter biochar was produced by pyrolyzing the chicken litter in an auger reactor at a temperature of 450°C and a holding time period of 20min [27]. The chicken litter also known as poultry litter consists of wood shavings, chicken feathers, chicken manure, bedding material *etc.* Laila *et al.* synthesized biochar from animal waste bones by pyrolysis which includes chicken, mutton, and beef bones, used as a fertilizer. The pyrolysis process was carried out in a locally assembled semi-automatic pyrolyzer. Biochar was obtained by heating animal bones under the absence of oxygen to a temperature of 550°C with a ramping rate of 30°C min^{-1} in the semi-automatic pyrolyzer [28]. Niu *et al.* synthesized biochar by using a carbonization process from cattle bone powder. The cattle bone powder was heated to temperatures of 600°C, 850°C, 900°C, 1000°C, 1100°C, and 1200°C respectively in an Argon atmosphere at a heating rate of 2.5 °C min-1, and kept at that temperature for 1h in the furnace. The generated biochar is first rinsed with HCl and deionized water before being dried at 80°C for 12 hours to create activated carbon [18]. Anil *et al.* [29] synthesized biocarbon of eggshells by carbonization process. For this, dried eggshell pieces were placed in a muffle furnace, heated to a temperature of 650°C and kept at that temperature for 1hr. Similarly, Kowshik *et al.* [30] prepared a carbonized eggshell powder by keeping the eggshells and graphite powder in a graphite crucible and heating them in a muffle furnace to a temperature of 850°C up to 1hr. A hydrothermal conversion process is also used for the synthesis of carbon dots which are zero-dimensional fluorescent nanomaterials. Hydrothermal carbonization is a thermochemical process, that typically takes place in pressurized water at a relatively low temperature (between 100 and 240 degrees Celsius), and transforms biomass precursors into solid biochar or structured carbon [21]. Ashraf *et al.* [31] synthesized carbon dots (CD) from prawn shells. 1.5gm of prawn shells and 30ml of the glacial acetic acid solution are taken in a Teflon lined hydrothermal reactor, heated to 180°C with a residence time of 7hrs. The liquid is then filtered to produce luminous CDs once the furnace has been allowed to cool to ambient temperature. There has been growing interest in exploring alternatives to traditional pyrolysis, such as microwave pyrolysis and solar hydrothermal carbonization, which feature even more heat distribution throughout the system, as well as laser and plasma pyrolysis, which are still too expensive for widespread use in the large-scale production of biocarbon materials [32].

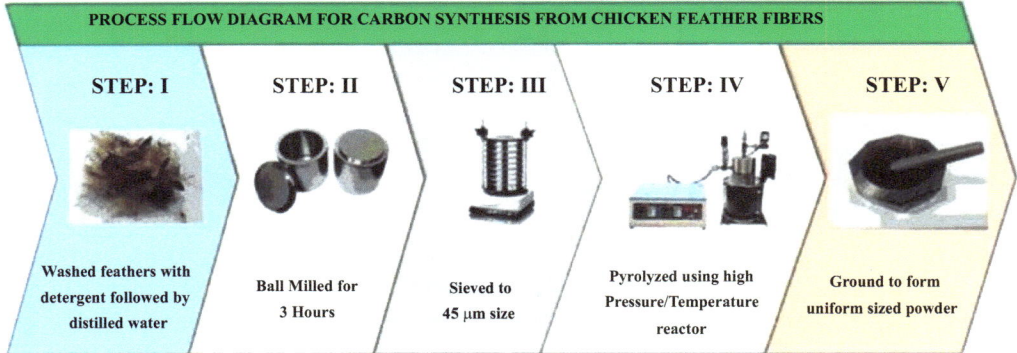

Fig. (1). Process flow diagram for carbon synthesis from chicken feathers [26].

3. POLYMER COMPOSITE FABRICATION

The selection of polymer has a decisive influence on the fabrication of polymer composites as this polymer is a matrix phase that protects the surface of the reinforcement from abrasion, chemical attack, and moisture absorption. Matrix materials for biochar composites can be either thermoplastic or thermoset polymers. Polymer composite with biocarbon reinforcement can be prepared by melt blending process, solution casting, and In-situ polymerization. If the matrix material is a thermoset polymer the composite is prepared by hand lay-up process, injection moulding, compression moulding methods, *etc.* Li *et al.* [24] prepared a polymer composite by combining PLA with 20 wt% of biocarbon obtained by pyrolysis of Chicken feather (ChF) and ChF powder at 300°C and 600°C using a twin-screw extruder. Rodriguez *et al.* [25] also prepared a polymer composite by combining polypropylene with 27wt% of biocarbon derived from ChF and shrimp shells. Mohammed *et al.* [26] prepared a supercapacitor by solution casting and 3D printing. The solid electrolyte layer of a supercapacitor is prepared by solution casting method by mixing polyvinyl alcohol (PVA) with phosphoric acid (H_3PO_4). Carbon paste was prepared by mixing activated carbon with PVA/H3PO4 electrolyte solution. This carbon paste was injected into the 3D printer's head through a 10 cc syringe fitted with a 0.8 mm nozzle. The carbon paste electrodes for the supercapacitor were printed using a Hyrel30M printer fitted with an SDS-10 syringe head. Anil *et al.* [29] in his experiment fabricated a thermoset polymer composite by using hand lay-up and compression moulding techniques. Hybrid composites were prepared by reinforcing chopped Borassus flabellifer fine fibres (BFF fibres) and carbonized eggshell powder in an Isophthalic polyester matrix. Carbonized eggshell powder is sonicated with matrix using ultrasonication to avoid agglomeration before preparing the composite. Kowshik *et al.* [30] also prepared a hybrid composite by reinforcing carbonized eggshell powder and E-glass fibre mats in polyester resin as shown in Fig. (**2**). He also prepared a hybrid composite by reinforcing uncarbonized and carbonized eggshell powder and glass fibre mat in a polyester resin. Das *et al.* [27] prepared a polymer composite by reinforcing 24wt% of carbon synthesized from poultry litter and 30% of wood in 42% of polypropylene. The composite was fabricated by extrusion/melt blending using a twin-screw extruder. The extrusion was carried out by operating the screw at 90 RPM and the temperature from the barrel to the die zone was adjusted between 175°C and 200°C. Epoxy polymer composite coating was prepared by reinforcing carbon dots synthesized from prawn shells in a Bisphenol (Bisphenoldiglycidylethe) epoxy polymer and polypropylene glycol bis (2-aminopropyl ether) hardener.

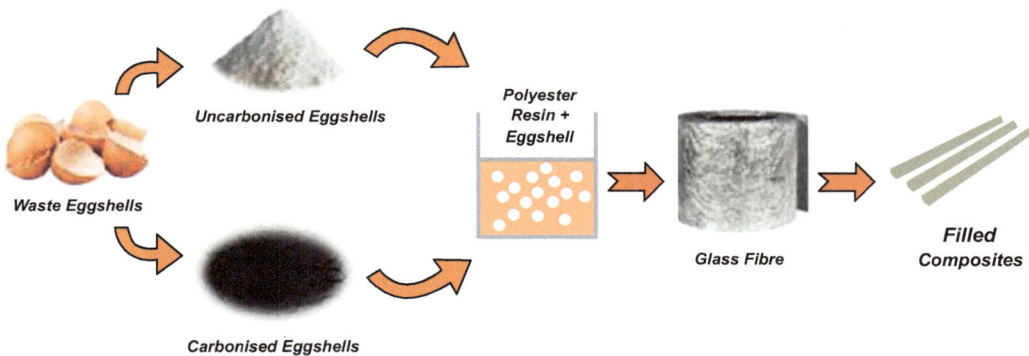

Fig. (2). Polymer composite preparation from carbonized eggshell [30].

4. THERMAL AND MECHANICAL PROPERTIES OF POLYMER COMPOSITES

The study of the characteristics of animal-based biocarbon-reinforced polymer composites is very crucial. In biocarbon-reinforced composites, the properties of the biocarbon are very important. The characteristics of biocarbon that affect the properties of biocarbon-reinforced composites are the type of biocarbon, size of the particles, loadings of biocarbon, and preparation temperatures of the biocarbon. In addition, the kind of polymer also has a significant impact on biocarbon-reinforced polymer composites. In this section, the mechanical and thermal properties of animal-based biocarbon-induced polymer composites are discussed.

4.1. Mechanical Properties

Mechanical properties play a crucial role in assessing the quality of biocarbon-reinforced composites. By performing a tensile test and a flexural test, the mechanical properties of the biocarbon-induced composites were evaluated. As these properties represent the bio composite's ability to support loads, they are the most important metrics to consider when evaluating them for applications. From this perspective, numerous studies have evaluated and reported on the mechanical characteristics of biochar composites. Studies have shown that with the addition of small amounts of biocarbon, there is a considerable improvement in mechanical properties. An improvement in the mechanical properties of the composite can be observed by adding 24 wt% biocarbon particles of poultry litter along with 30% of wood in a polypropylene matrix. The tensile strength and tensile modulus of the composite are observed as 27MPa and 4.28GPa, respectively which is more than neat polymer [27]. Similar results were observed in hybrid composites of glass fibre/polyester reinforced with carbonized eggshell powder as filler. The tensile and flexural strengths of carbonized eggshell-filled composites were 43.9 and 34.14% greater than those of unfilled composites, respectively. This increase in tensile strength is due to the larger surface area of the carbonated eggshell particles which transfers tensile stress to the filler from the matrix [30]. Similar outcomes were obtained when carbonized eggshell powder and constant amounts of treated Borassus flabellifer fibres were incorporated into the Isophthalic polyester matrix. The filler content is varied from 1wt% to 3wt%. With the addition of 2 wt% biocarbon, an improvement of 121% in tensile strength, 76% in flexural strength, and 113% in impact strength was observed [29]. When filler loading is further increased, there is deterioration in the mechanical properties of the composite [25]. This might occur as a result of greater filler loading, causing non-homogeneous filler particle dispersion, which results in the agglomeration of the filler particles. The decrement in mechanical properties was also observed by Li *et al.* [24] in their experimentation by adding chicken feather biocarbon in a polymer matrix. The tensile strength, elongation at break, and flexural strength decreased with the addition of 20 wt% of biocarbon for all the samples compared with the neat polymer due to the agglomeration of carbon particles. As the pyrolysis temperature increases, the tensile strength and flexural strength of the composite are increased. The tensile strength and flexural strength of the 600°C sample increased by 31 and 53% than the 300°C sample, which is shown in Fig. (**3**).

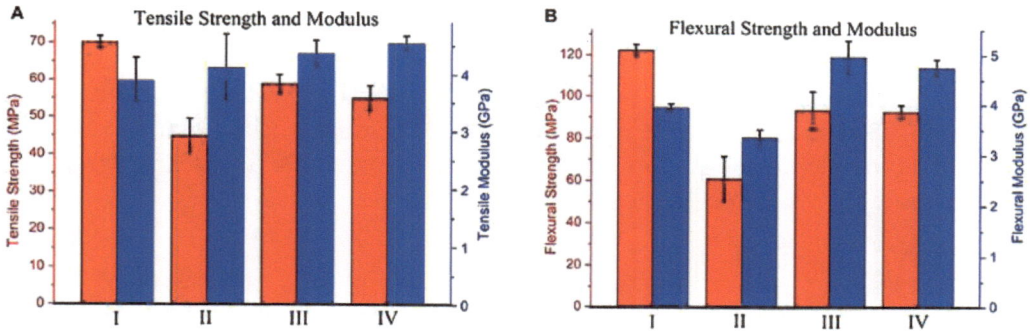

Fig. (3). Bar graph representing A) ChF biocarbon tensile strength and tensile modulus B) ChF biocarbon Flexural strength and Flexural modulus of (I) Neat PLA, (II) 80/20 PLA/ biocarbon at 300°C (III) 80/20 PLA/biocarbon at 600°C (IV) 77/20/3 PLA/biocarbon/MA-g-PLA 600°C [24].

This is due to low interfacial adhesion. This low interfacial adhesion is due to low pyrolysis temperature and reduction of the size of the particle. To improve this, interfacial adhesion and compatibilizers can be added. These compatibilizers reduce the gaps and improve the tensile strength.

4.2. Thermal Properties

Thermal analyses, such as thermogravimetric analysis (TGA) and differential scanning calorimeter analysis (DSC), were accomplished to describe the thermal characteristics to satisfy the requirements for high-quality bio-composites. According to the DSC analysis, adding biocarbon caused a little increase in the crystallization temperature (Tc), suggesting that the biocarbon might act as a nucleating agent for the polymer. Also, with the inclusion of biochar, the decomposition temperature shifted to a higher value than the neat polymer [27]. However, the thermal stability of the samples was marginally decreased due to the incorporation of biocarbon [24].

CONCLUSION

Developing biochar from animal waste can be an effective way of using the millions of tons of animal waste generated each year. As a result of its widespread applicability and promising future, biochar has emerged as a promising candidate for use in contemporary scientific and technological endeavours. Therefore, this review summarizes the various sources of animal wastes, synthesizing methods of biochar, fabrication of polymer composites, and thermomechanical characteristics of animal-based biochar-reinforced polymer composites. It can be concluded from this review that the inclusion of biochar as a reinforcing filler in polymer composites enhances mechanical and thermal properties with minimum loading of biochar. Various efforts have been made to synthesize biocarbon from animal waste and use it as an electrode material, fertilizer, and adsorbent. However, more research is needed to develop polymer nanocomposites as these composites exhibit good thermal, mechanical, and electrical properties.

REFERENCES

[1]   Thippeswamy, B.H.; Maligi, A.S.; Hegde, G. Roadmap of effects of biowaste-synthesized carbon nanomaterials on carbon nano-reinforced composites. *Catalysts,* **2021**, *11*(12), 1485.
[http://dx.doi.org/10.3390/catal11121485]

[2]   Khan, M.B.; Cui, X.; Jilani, G.; Tang, L.; Lu, M.; Cao, X.; Sahito, Z.A.; Hamid, Y.; Hussain, B.; Yang, X.; He, Z. New insight into the impact of biochar during vermi-stabilization of divergent biowastes: Literature synthesis and research pursuits. *Chemosphere,* **2020**, *238*, 124679.
[http://dx.doi.org/10.1016/j.chemosphere.2019.124679] [PMID: 31524617]

[3]   Fang, Z.; Gao, Y.; Bolan, N.; Shaheen, S.M.; Xu, S.; Wu, X.; Xu, X.; Hu, H.; Lin, J.; Zhang, F.; Li, J.; Rinklebe, J.; Wang, H. Conversion of biological solid waste to graphene-containing biochar for water remediation: A critical review. *Chem. Eng. J.,* **2020**, *390*(February), 124611.
[http://dx.doi.org/10.1016/j.cej.2020.124611]

[4]   Yahya, M.A.; Al-Qodah, Z.; Ngah, C.W.Z. Agricultural bio-waste materials as potential sustainable precursors used for activated carbon production: A review. *Renew. Sustain. Energy Rev.,* **2015**, *46*, 218-235.
[http://dx.doi.org/10.1016/j.rser.2015.02.051]

[5]   Ciencia, M.; Issn, O. Gacitua, W.; Ballerini, A.; Zhang, J. Polymer nanocomposites: synthetic and natural fillers a review. *Maderas Ciencia Y Tecnología*, **2005**, *7*(3), 159-78.

[6]   Nanotubes, C.; Nanofillers, C. "Polymer Nanocomposites — A Comparison between," no. *Figure,* **2016**, *1*, 1-35.
[http://dx.doi.org/10.3390/ma9040262]

[7]   Richard, S.; Rajadurai, J.S.; Manikandan, V. Influence of particle size and particle loading on mechanical and dielectric properties of biochar particulate-reinforced polymer nanocomposites. *IJPAC Int. J. Polym. Anal. Charact.,* **2016**, *21*(6), 462-477.
[http://dx.doi.org/10.1080/1023666X.2016.1168602]

[8]   Santhiago, M.; Garcia, P.S.; Strauss, M. Bio-based nanostructured carbons toward sustainable technologies. *Curr. Opin. Green Sustain. Chem.,* **2018**, *12*, 22-26.
[http://dx.doi.org/10.1016/j.cogsc.2018.04.009]

[9]   Nasir, S.; Hussein, M.; Zainal, Z.; Yusof, N. Carbon-based nanomaterials/allotropes: A glimpse of their synthesis, properties and some applications. *Materials (Basel),* **2018**, *11*(2), 295.
[http://dx.doi.org/10.3390/ma11020295] [PMID: 29438327]

[10]   Inagaki, M.; Kang, F. Introduction. In: Inagaki. M. Materials Science and Engineering of Carbon: Characterization. Elsevier, Butterworth-Heinemann, **2016**: pp. 1-6.
[http://dx.doi.org/10.1016/B978-0-12-805256-3.00001-5]

[11]   Devi, N. S.; Hariram, M.; Vivekanandhan, S. Modification techniques to improve the capacitive performance of biocarbon materials *J. Energy Storage,* **2021**, *33*.
[http://dx.doi.org/10.1016/j.est.2020.101870]

[12]   Speranza, G. The Role of Functionalization in the Applications of Carbon Materials: An Overview. *C — Journal of Carbon Research,* **2019**, *5*(4), 84.

[http://dx.doi.org/10.3390/c5040084]

[13]   Demirbas, A. Effects of temperature and particle size on bio-char yield from pyrolysis of agricultural residues *J. Anal. Appl. Pyrolysis,* **2004**, *72*, 243-248.
[http://dx.doi.org/10.1016/j.jaap.2004.07.003]

[14]   S. S. SBiochar from biomass waste as a renewable carbon material for climate change mitigation in reducing greenhouse gas emissions — a review , **2020**; 280, .

[15]   Riuji, C.; Stefan, L.; Imanol, D.; Mertenat, A.; Zurbru, C. Treatment technologies for urban solid biowaste to create value products: A review with focus on low- and middle- income settings , **2017**; pp. 81-130.
[http://dx.doi.org/10.1007/s11157-017-9422-5]

[16]   Salim, N.V.; Jin, X.; Mateti, S.; Lin, H.; Glattauer, V.; Fox, B.; Ramshaw, J.A.M. Porous carbon fibers made from collagen derived from an animal by-product. *Materials Today Advances,* **2019**, *1*, 100005.
[http://dx.doi.org/10.1016/j.mtadv.2019.100005]

[17]   Khalil, M.; Berawi, M.A.; Heryanto, R.; Rizalie, A. Waste to energy technology: The potential of sustainable biogas production from animal waste in Indonesia. *Renew. Sustain. Energy Rev.,* **2019**, *105*(February), 323-331.
[http://dx.doi.org/10.1016/j.rser.2019.02.011]

[18]   Niu, J.; Shao, R.; Liang, J.; Dou, M.; Li, Z.; Huang, Y.; Wang, F. Biomass-derived mesopore-dominant porous carbons with large specific surface area and high defect density as high performance electrode materials for Li-ion batteries and supercapacitors. *Nano Energy,* **2017**, *36*, 322-330.
[http://dx.doi.org/10.1016/j.nanoen.2017.04.042]

[19]   Das, S.K.; Ghosh, G.K.; Avasthe, R. Applications of biomass derived biochar in modern science and technology. *Environmental Technology & Innovation,* **2021**, *21*, 101306.
[http://dx.doi.org/10.1016/j.eti.2020.101306]

[20]   Kim, H.S.; Kang, M.S.; Yoo, W.C. Co3O4 nanocrystals on crab shell-derived carbon nanofibers (Co3O4@CSCNS) for high-performance supercapacitors. *Bull. Korean Chem. Soc.,* **2018**, *39*(3), 327-334.
[http://dx.doi.org/10.1002/bkcs.11389]

[21]   Senthil, C.; Lee, C. W. Biomass-derived biochar materials as sustainable energy sources for electrochemical energy storage devices. *Renew. Sustain. Energy Rev,* **2020**, *137*(10), 110464.

[22]   Dhyani, V.; Bhaskar, T. A comprehensive review on the pyrolysis of lignocellulosic biomass. *Renew. Energy,* **2018**, *129*, 695-716.
[http://dx.doi.org/10.1016/j.renene.2017.04.035]

[23]   Czajczyńska, N.D.; Anguilano, L.; Ghazal, H.; Krzyżyńska, R.; Reynolds, A.J.; Spencer, H.J. Potential of Pyrolysis Processes in the Waste Management Sector. *Therm. Sci. Eng. Prog.,* **2017**, *3*, 171-197.

[24]   Li, Z.; Reimer, C.; Picard, M.; Mohanty, A.K.; Misra, M. Characterization of Chicken Feather Biocarbon for Use in Sustainable Biocomposites. *Front. Mater.,* **2020**, *7*(February), 3.
[http://dx.doi.org/10.3389/fmats.2020.00003]

[25]   Rodriguez-Uribe, A., Tripathi, N., Mohanty, A.K., Misra, M. Biocarbon Produced from Agricultural, Industrial, Forestry, and Food Waste: Performance Evaluation of Biocarbon-Reinforced Polypropylene Biocomposites. Society of Plastics Engineer's (SPE) Automotive Composites Conference & Exhibition (ACCE), Novi, USA. September 7-9, **2022**.

[26]   Mohammed, Z.; Jeelani, S.; Korivi, N. S.; Rangari, V. Synthesis and characterization of N-doped porous carbon from chicken feathers for 3D printed electrode applications. **2022**. CC BY 4.0.
[http://dx.doi.org/10.21203/rs.3.rs-2230929/v1]

[27]   Das, O.; Sarmah, A.K.; Bhattacharyya, D. Biocomposites from waste derived biochars: Mechanical, thermal, chemical, and morphological properties. *Waste Manag.,* **2016**, *49*, 560-570.

[http://dx.doi.org/10.1016/j.wasman.2015.12.007] [PMID: 26724232]

[28] Hussain, A.; Nazir, A.; Shafiq, M. Potential Application of Biochar Composite Derived from Rice Straw and Animal Bones to Improve Plant Growth. *Sustain,* **2021**, *13*(19), 11104.

[29] Baby, A.; Nayak, S.Y.; Heckadka, S.S.; Purohit, S.; Bhagat, K.K.; Thomas, L.G. Mechanical and Morphological Characterization of Carbonized Egg-shell Fillers/ Borassus Fibre Reinforced Polyester Hybrid Composites. *Mater. Res. Express,* **2019**, *6*(10), 105342.

[30] Kowshik, S.; Sharma, S.; U, S.R.; Shettar, M.; Hiremath, P.; Upadhyaya, A. Investigation on the Effects of Uncarbonised, Carbonised and Hybrid Eggshell Filler Addition on the Mechanical Properties of Glass Fibre/Polyester Composites. *Engineered Science,* **2022**, *18*, 121-131. [http://dx.doi.org/10.30919/es8d679]

[31] Ashraf, P.M.; Anju, V.S.; Binsi, P.K.; Joseph, T.C. A green extraction process of nanocarbon dots from prawn shells, and its reinforcement in epoxy polymers. *J. Appl. Polym. Sci.,* **2023**, *140*(1), e53250. [http://dx.doi.org/10.1002/app.53250]

[32] Alhelal, A.; Mohammed, Z.; Jeelani, S.; Rangari, V.K. 3D printing of spent coffee ground derived biochar reinforced epoxy composites. *J. Compos. Mater.,* **2021**, *55*(25), 3651-3660. [http://dx.doi.org/10.1177/00219983211002237]

[33] Fu, M.; Chen, W.; Ding, J.; Zhu, X.; Liu, Q. Biomass waste derived multi-hierarchical porous carbon combined with CoFe2O4 as advanced electrode materials for supercapacitors. *J. Alloys Compd.,* **2019**, *782*, 952-960. [http://dx.doi.org/10.1016/j.jallcom.2018.12.244]

[34] Côrtes, L.N.; Druzian, S.P.; Streit, A.F.M.; Sant'anna Cadaval Junior, T.R.; Collazzo, G.C.; Dotto, G.L. Preparation of carbonaceous materials from pyrolysis of chicken bones and its application for fuchsine adsorption. *Environ. Sci. Pollut. Res. Int.,* **2019**, *26*(28), 28574-28583. [http://dx.doi.org/10.1007/s11356-018-3679-2] [PMID: 30446910]

[35] Ou, J.; Zhang, Y.; Chen, L.; Zhao, Q.; Meng, Y.; Guo, Y.; Xiao, D. Nitrogen-rich porous carbon derived from biomass as a high performance anode material for lithium ion batteries. *J. Mater. Chem. A Mater. Energy Sustain.,* **2015**, *3*(12), 6534-6541. [http://dx.doi.org/10.1039/C4TA06614F]

CHAPTER 4

Harnessing Agro-based Biomass for Sustainable Thermal Energy Storage with Biochar Polymer Nanocomposites

Venkateswara Rao Kode[1,*]

[1] *Department of Chemical and Biochemical Engineering, Christian Brothers University, Memphis, TN 38104, USA*

Abstract: Food loss due to wastage is a severe issue for the entire world. Wastage accounts for up to one-third of lost food output globally. Although there are various ways to turn food waste into useful functional materials, landfilling is still a frequent practice. This results in greenhouse gas emissions, which increase the already significant greenhouse gas emissions linked to the agriculture sector. In this review, we start by identifying multiple biomass sources and synthesis methods for biochar made from biomass. This contains methods for the characterization of phase change polymer nanocomposite materials impregnated with biochar. In order to compare the thermal properties of phase change materials and gain an understanding of the various biowastes, a comprehensive methodology was used.

Keywords: Biomass, Biochar, Composite Phase Change Materials, Energy Storage, Sustainability.

1. INTRODUCTION

Agricultural waste disposal continues to be a major socio-economic problem. As rapid urbanization and population growth continue to mount pressure on the manufacturing and agricultural industries to meet the demand, waste generation will accelerate. Agricultural waste, such as biomass residues, is an underutilized feedstock for synthesizing valuable products. More than 140 gigatons of biomass waste are produced globally every year [1], providing a considerable stream of feedstock material. Biomass disposal has a devastating effect on our environment.

Although the repercussions led by biomass disposal are alarming, this issue presents an opportunity to utilize this ample material as a potential feedstock.

[*] **Corresponding author Venkateswara Rao Kode:** Department of Chemical and Biochemical Engineering, Christian Brothers University, Memphis, TN 38104, USA; E-mail: vkode@cbu.edu

Deepa Kodali & Vijaya Rangari (Eds.)

Biomass is an abundant renewable resource, as the supply can be replenished over a short amount of time and is virtually limitless [2]. In the United States, roughly 386 million tons of biomass waste is generated annually. The agriculture industry remains a vital part of the US economy, providing 10.3% of the nation's employment and 5.2% of the country's overall GDP [1]. Biomass refers to any organic material that comes from living or recently living organisms. This can include plants, crops, trees, and agricultural and forestry residues, as well as waste from the food, paper, and biofuel industries. A good example of this is almond biomass, wherein, over the past several years, the demand for almond-based products has skyrocketed, resulting in over three billion pounds of almond biomass waste being produced annually. The global production of almonds has significantly grown from ≈0.7 million tons in 2007 to ≈1.5 million tons in 2019 [1], enabling the simultaneous accumulation of almond woody biomass, which can negatively impact the environment and economics of almond production. More importantly, when producing almond kernel for human consumption, the remaining almond hulls, shells, and woody biomass (*e.g.*, twigs and wood clippings) from the production process do not have much commercial value.

According to the International Nuts and Dried Fruits Statistical Yearbook, the USA is the largest producer of almonds, with over 700,000 hectares of land dedicated to almond production [2]. For instance, California alone produces 80% of the almonds used worldwide, and production is rising. Almonds are grown and processed by roughly 7600 people annually, yielding about 746,000 tons of shells. More specifically, the average production in California in 2019 included 1.03 million tons of kernels, 0.72 million tons of shells, and 2.06 million tons of hulls [3, 4]. On the other hand, walnut shells are useful for producing biochar due to their easily modifiable pore structure and volatile nature. Walnuts are a commonly produced nut, with China producing 1.06 million tons out of the 3.7 million tons produced globally in 2019. Moreover, the shells of walnuts account for 60% of the fruit's mass [5].

Historically, biomass waste has mostly been either burnt or disposed of due to mobility issues and, thus, the lack of adequate applications. While some biomass is repurposed into sources of energy or heat, the majority is disposed of in incinerators and landfills, accounting for 18% of total global emissions of CO_2 every year and resulting in deteriorated surface and groundwater quality [1]. These are often used as bedding for poultry. This leads to an increase in greenhouse gas emissions and other environmental pollution. Furthermore, almond woody biomass left in the field as mulch has been shown to cause reduced nitrogen availability, which could affect crop yields and productivity.

Thus, new, more environmentally friendly solutions are needed to minimize environmental hazards. One of the promising pathways of waste management is utilizing them in developing bio-based polymer composites, which are eco-friendly materials. Especially, biomass-derived biochar and their subsequent fabrication of phase change composite materials are interesting for thermal energy storage applications.

Hence, this book chapter reviews synthesizing biochar from different biomass feedstock, including almond and walnut shells. This includes techniques for biochar impregnation of phase change materials, and characterization. Finally, a holistic approach is undertaken to compare the thermal properties of phase change materials to gain insight into the diverse biowastes.

2. COMPOSITION

2.1. Almonds

The almond tree, one of the most popular nut-based trees for snack foods and ingredients in many processed foods, yields almonds. The International Nut and Dried Fruit Council (INC) reports that in 2017 and 2018, almost 1.2 million tons of almonds were produced. The United States leads the world in production with 81% of the total. With 7% and 4%, Australia and Spain are close behind, respectively.

The hull and shell enclosing the edible kernel are mechanically isolated to produce the almond biomass (Fig. **1**) during the hulling and shelling process [6]. Almond hulls, which make up 52% of the fresh fruit, have erratic ripening cycles that change depending on several growing conditions [7]. The intermediate shell, which accounts for around 33% of the total fresh weight, can house a variety of shapes and sizes with varying mechanical properties that are greatly influenced by environmental factors [6].

Fig. (1). Photograph of three different as-received almond woody biomass including hulls, shells and twigs.

Exploring almond byproducts as a potential source of high-value phase change composite materials is essential in order to improve the sustainability and competitiveness of the almond business while reducing negative environmental effects. However, it is crucial to comprehend the makeup of almond waste, in particular the hulls, shells, and twigs, which are mostly made of cellulose, hemicellulose, and pectin, as well as lignin, tannin-like poly-phenolic compounds, and ash [8]. Quantifying the almond biomass's composition is therefore crucial for the effective production of biochar and its derivatives for uses in thermal energy storage.

Li *et al.* examined the physical structure and chemical properties of almond shells. Almond shells have holes with widths ranging from 40 to 60 mm for microscopic holes and 300 to 500 mm for bigger ones, according to a microscope examination. Almond shells contain the following elements in that order: C, O, N, and Si (72.72%, 22.88%, 3.87%, and 0.87%, respectively). Almond shells contain 38.47% cellulose, which means that a composite constructed of almond shells may have better mechanical characteristics than most biomass nut shells. The three main chemical components are cellulose, hemicellulose, and lignin for 38.48%, 28.82%, and 29.54%, respectively.

2.2. Walnuts

There are more than 3.7 million tons of walnuts produced worldwide. With approximately 1.06 million tons produced, China is the largest producer, followed by Turkey, Iran, and the United States, in that order. Portugal produced 4,000 tons of walnuts [9]. The walnut shell contains 49.12% lignin, 31.19% cellulose, and 16.59% hemicellulose, according to a research by Yuan *et al.* [7]

3. BIOCHAR

As a carbon-rich substance known as biochar, biomass can be thermochemically converted in an oxygen-limited environment. It has a fine-grained, porous powder consistency. The market value of the biochar sector is anticipated to increase significantly over the next ten years, according to historical data. Transparency Market Research did a study in an effort to forecast worldwide biochar market revenue. It predicted that the market would expand at a Compound Annual Growth Rate (CAGR) of 15.35% from 2017 to 2031, reaching a value of more than $6.3 billion. They projected the following decade using the data available to 2020. According to their extensive study report, the worldwide biochar industry, which was estimated to be worth $1320.94 million in 2021, will have a staggering $3171.02 million in sales after eight years. Volumetrically, the market share of 394.09 kilotons would have amounted to 781.09 kiloton during the same time frame. Supporting the expansion of the biochar market is thought to have various benefits for the environment. With a remarkable revenue growth rate of 13.40% in revenue and 10.34% in volume, it offers the stockholders a lot of potential.

3.1. Synthesis

The protocol undertaken to produce biochar is a dominant factor in determining the physiochemical characteristics of the prepared biochar, in addition to the nature of the starting material. If the temperature of the process implemented is chosen as the classifying parameter, we have techniques like slow and fast pyrolysis with a temperature range of 400-600°C. At temperatures below this range, a process called hydrothermal carbonization occurs (180-250°C) while microwave gasification occurs at much higher temperatures of 800-1000°C. As stated earlier, the phase of the end product is a function of the process implemented. The end product is solid if pyrolysis, dry carbonization or gasification is executed while hydrothermal carbonization leads to a slurry form of the biomass if high pressure is used.

In terms of the efficiency of the process, hydrothermal carbonization is the best of the available methods of preparation. With a yield of 30-60%, it is a superior process compared to the 10% char yield of microwave gasification. However, it has the disadvantage of being a time-consuming process. While it takes 1-12 hours for the whole process to be completed, the latter process takes a mere 5-20 seconds. Fast pyrolysis takes a mere 1 second and has a yield of 10-20% as opposed to the 20-40% yield obtained from slow pyrolysis. The residence time of slow pyrolysis is the range of minutes to days.

3.2. Pyrolysis

It is the approach used to create biochar most frequently. The slow version of the process takes up to several days and has been the method used since centuries because the low heating rate is economically feasible with respect to the energy input. It is alternatively called conventional carbonization. The low temperature conditions are feasible for biological processes to occur, but the long residence time proves to be a challenging task to sustain the process up until fruition. The fast pyrolysis method does not have this issue as its residence time is less than 10 seconds and the rate of heating is as high as 200 K/min.

The side products created during the process have high calorific value too. They are usually obtained in liquid and gaseous forms. Pyrolysis carried out in the absence of oxygen results in carbon-rich products that are more resistant to microbial degradation compared to the starting material due to the change in carbon constituents. This is because the oxygen-free environment changes the carbon constituents of the biomass. In contrast, thermochemical conversion leads to three different end products, all having a variety of uses. These products are bio-oil, biochar and syngas. The latter two do not require any processing before they can be used. One of the main differences between the two types of methods discussed is that slow pyrolysis has a higher biochar yield while slow pyrolysis has a higher bio-oil yield.

3.3. Gasification

As implied by the name, this process involves turning the initial source material (biomass) into primarily gaseous products including carbon monoxide, carbon dioxide, methane, hydrogen, and minor amounts of higher hydrocarbons. Syngas is the name of their mixture, which has numerous industrial uses. Oxygen, air, steam, or mixes of these gases are provided in a controlled quantity, and the entire process takes place in a controlled setting at temperatures as high as 700°C. Syngas produced using steam as the oxidizing agent has a higher heating value of 10-14 MJ/Nm^3, while that produced using air has a lower heating value of 4-7 MJ/Nm^3. About 10% of the biomass feedstocks are typically converted into biochar during gasification.

3.4. Hydrothermal Carbonization

Abbreviated as HTC, this process requires water above 160-800 °C as a medium. To reduce the boiling point of water and prevent its spontaneous conversion to steam, a pressure of more than 1 atm needs to be maintained.

There are two main classifications of HTC: high temperature HTC (300-800 °C) and low temperature HTC (<300 °C). Hydrothermal gasification, resulting in gases like hydrogen and methane, dominates the former process as the high temperatures do not support the stability of most of the organic compounds. At slightly lower temperatures, carbonization of biomass is the dominant process.

Low temperature HTC is similar to the conversion of biomass to coal as it takes place in nature. It has a higher reaction rate and a shorter reaction time though. The yields vary from 30-60% and depend on the reaction conditions and properties of the feedstock. It is an economical process of biochar production for methods requiring a high moisture environment since HTC medium is water.

4. BIOCHAR-ENCAPSULATED POLYMER NANOCOMPOSITES:

4.1. Synthesis of Phase Change Materials

Fig. **(2)** illustrates the preparation of polyethylene glycol (PEG) impregnated with almond shell biochar phase change composite materials [8]. PEG mass fractions were selected in the range of 40-70%. PEG dissolved in ethanol solution was added to the almond shell biochar powder. The samples were then vacuum-impregnated under room temperature while stirring the sample for 1.5 hours at 500r/min. Later, the samples were bath sonicated for 6 hours at 50 °C with continued vigorous stirring at 500r/min. The resulting composite was dried for 24 hours at 40 °C for subsequent quality testing and material characterization.

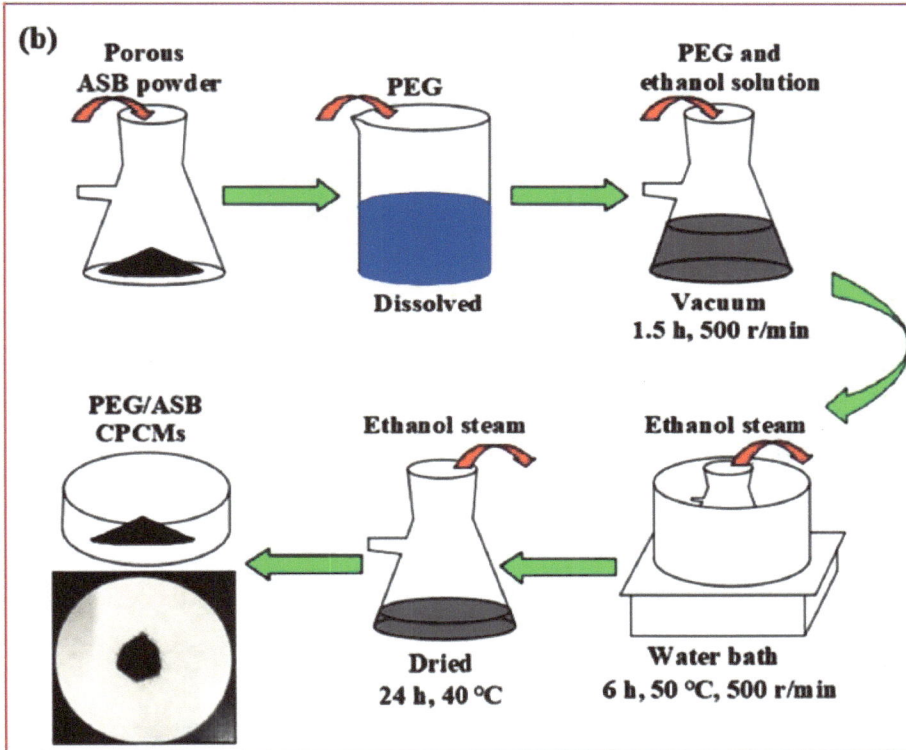

Fig. (2). Schematic illustration of PEG/almond shell biochar phase change composite material preparation. [reprinted with permission from reference [8]].

4.2. Characterization

Recently, there has been a growing interest in using biomass to create advanced functional materials, thanks to scientific and technological advancements. Biochar, which has strong adsorption ability and high thermal and electrical conductivity, has gained recognition as a promising material for creating phase change materials that can be used in heating and cooling applications by absorbing and releasing energy at the phase transition. By incorporating biochar, phase change materials can also exhibit powerful photothermal and electrothermal conversion properties. With the unique features of biochar, it is now possible to develop sophisticated form-stable composite materials for various energy storage applications. However, to maximize the efficiency of using biochar as a composite material for thermal energy storage, it is important to understand the surface morphology of the synthesized biochar, including factors such as pore distribution, surface area, and surface functionality.

Researchers who are interested in biochar-engineered phase transition material encapsulation and thermal energy storage have already reported on several experiments. The relation between the characteristics of biochar and how they affect improving performance in phase change composite materials still needs to be clarified, though. No clear correlation between the properties and process circumstances has been found, despite numerous articles reporting connections between biochar properties and their respective production and impregnation conditions. To optimize the properties of biochar-based form-stable composite materials for their efficacy in any application, research on characterizations of biochar properties and their relationship to phase change materials utilized for its synthesis is essential.

For example, Chen *et al.* [8] developed an environmentally friendly, sustainable, and cost-effective biochar from almond shell feedstock using a simple pyrolysis technique. Earlier, almond shells that were procured from an apricot plantation in China were cleaned, dried at elevated temperatures, and crushed to powder. Then, for the creation of biochar, the particles with sizes lower than 0.074 mm were gathered. To create a new form-stable phase change composite material, PEG was impregnated into the synthesized biochar by vacuum, which had the appropriate pore structures.

The SEM pictures (Fig. **3a**) of biochar produced by pyrolyzing almond shell biomass revealed significant porosity and surface area, which is a crucial characteristic for the PEG adsorption and energy storage. The pore size distribution further confirmed that almond shell biochar has a total pore volume of 0.17 cm^3/g and a surface area of 291.21 m^2/g, suggesting that almond shell biochar may possess favorable adsorption ability. Notably, the average pore diameter determined by the pore size distribution curve was estimated to be 2.33 nm, indicating that the almond shell biochar exhibited mesoporous solid material properties optimal for phase change material impregnation [9]. It is to be noted that pyrolysis conditions influence the average pore size distribution. SEM images (Fig. **3b**) of phase change composite material that is almond shell biochar impregnated with 60 wt.% of PEG content further revealed that the pores are filled with PEG, indicating that the biochar is a reliable adsorbent to achieve efficient form-stable phase change composite material encapsulation. The Fourier transform infrared (FTIR) spectra further revealed that the interaction between almond shell biochar and PEG occurred through a simple physical process, including capillary and surface tension effects and van der Waals forces. The interaction phenomenon was also consistent with other studies reported by Wan *et al.* [10] and Hekimoglu *et al.* [11] between biomass-derived biochar and phase change material, where the interactions were mainly physical phenomena.

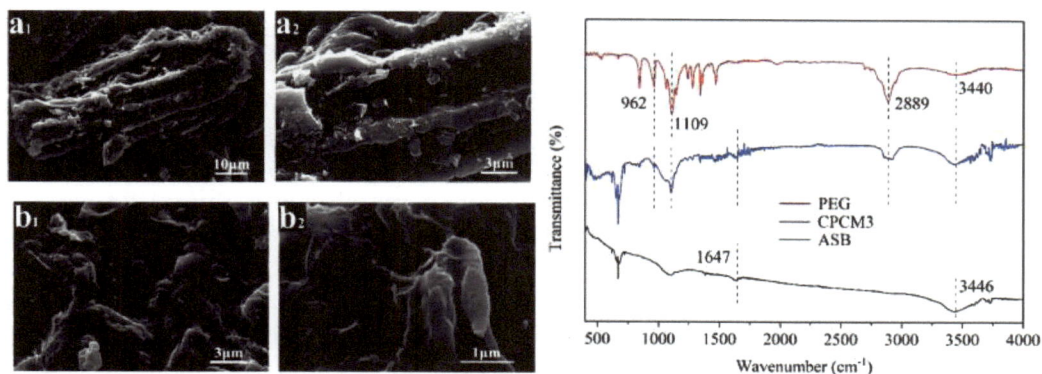

Fig. (3). SEM images of almond shell-derived biochar before and after impregnation of 60 wt.% PEG. FTIR spectra of PEG, almond shell biochar, and composite phase change material. [reprinted with permission from reference [8]].

A novel carbonaceous material was reported by Hekimoglu *et al.* in a recent study [11] using the walnut shell as a feedstock for carbonization or chemical activation. Briefly, as-received walnut shells were thoroughly washed with deionized water and dried at a higher temperature of up to 105 °C. Subsequently, the crushed walnut shell powder was subjected to pyrolysis at 500 °C under an inert atmosphere. Hekimoglu *et al.* [11] also prepared another feedstock material via chemical activation of the walnut shell by adding walnut shells to 60 wt.% $ZnCl_2$ salt solution at a mass ratio of 1:2. The treated sample was further processed and subjected to carbonization treatment to prepare activated walnut shell-derived biochar as a new carbon filler.

One of the promising pathways to determine the optimal loading concentrations of phase change materials during the impregnation of biochar is the leakage test. These tests were performed at varied experimental conditions, such as temperature, to understand the mechanical and thermal stability of biochar doped with phase change materials. To obtain leak-free composite phase change materials, Hekimoglu *et al.* [8] conducted a leakage test by impregnating methyl palmitate (MP) separately into walnut shell biochar and activated walnut shell biochar. The results of the experiments demonstrated that the ideal mass fractions of MP in the composites of biochar and activated carbon biochar were 43 and 55%, respectively (Fig. **4**). Leakage tests were previously carried out on activated almond shell biochar and unactivated almond shell biochar saturated with PEG by Chen *et al.* [8] During the experiment, composite phase change material samples (CPCMs) comprised of different percentages of PEG and almond shell biochar were heated in an oven at a temperature above the melting point of PEG. It was observed that pure PEG melted completely at 65°C and was not stable in its form. However, there was no leakage of liquid PEG when the PEG content was 40% (CPCM1), 50% (CPCM2), or 60% (CPCM3). The higher mechanical stability of form-stable phase change materials could be due to the capillary and surface tension forces between the porous almond shell biochar structure and PEG molecules. Additionally, the biochar's porous features enable mechanical strength to limit molten PEG leaking. However, almond shell biochar impregnated with 70% PEG content (CPCM4) showed evident traces of liquid PEG after sample removal, indicating the PEG leakage due to the excessive adsorption capacity of almond shell biochar.

Fig. (4). Leakage test results of composite phase change materials doped with PEG and almond shell biochar (CPCMs). MP impregnated separately [reprinted with permission from reference [8]].

The X-Ray Diffraction (XRD) patterns from the same study further revealed that the impregnation of almond shell biochar and activated almond shell biochar did not influence the crystallinity of MP [11]. The crystal structure represents the position and intensity of the different peaks in the XRD technique, which obtains information on the crystal structure. The instrument and microstructural factors have the greatest influence on the widths and forms of diffraction peaks. Each distinct peak is attributable to the scattering from a particular collection of atoms arranged in parallel planes. It implies that the location of the diffraction peaks is determined by the separation between parallel atomic planes. Increased crystallinity is indicated by a narrower XRD, and *vice versa*. The appearance of a narrow and different XRD peak indicates more crystallinity and a similar lower XRD indicates a lower degree of crystallinity. The XRD patterns of WSC and AWSC exhibited no sharp diffraction peaks, suggesting the non-crystalline amorphous structural behavior. On the other hand, the crystalline peaks at 8.00°, 12.09°, 16.15°, 21.87°, 24.09° and 37.24°, exhibited by pure MP remained similar with minimal shifts in the XRD curves of leak-free phase change composites, including walnut shell biochar and activated walnut shell biochar impregnated with MP [11]. While the XRD study by Chen *et al.* [8] depicted the same phenomenon, it was found that the intensity of XRD peaks was lowered post-impregnation of PEG into almond shell biochar, which could be due to limited PEG crystalline growth caused by the pore structures of almond shell biochar.

5. THERMAL PROPERTIES

To improve the efficiency of latent heat storage, it is important to understand the thermal stability and properties of form-stable phase change materials. This can be evaluated through techniques such as thermogravimetric analysis and differential scanning calorimetry. Researchers have conducted numerous studies to investigate the thermal degradation stability, thermal conductivity, and thermal performance of these materials. By incorporating form-stable phase change materials into building components, it is possible to create structures that are energy-efficient, lightweight, and affordable.

Table 1 provides a comparison of the latent heat of fusion of different composites that have been developed using biochar-based composite phase transition materials. The fusion enthalpies of these composites can be affected by the impregnation ratios of the phase change material composition. Thermal cycle tests have demonstrated the reliable latent heat energy storage properties of these composites, and TGA analysis has shown their remarkable heat stability. For instance, a study by Wan *et al.* [10] found that compared to pure palmitic acid, the thermal conductivity of pinecone biochar impregnated with the acid was enhanced by 43.76%.

Table 1. The latent heats of fusion and thermal conductivity of composite PCMs prepared from various biomass feedstock.

Feedstock	PCM	Fusion enthalpy (J/g)	Increment of Thermal conductivity (%)	References
Oil palm kernel shell	Paraffin	57.3	-	[12]
Corn cob	Lauric-stearic acid (LSA)	148.3	93.4	[13]
Pepper straw	palmitic acid (PA)	95.5	-	[14]
Coconut shell	PEG	81.3	-	[15]
Palm kernel shell	n-octadecane	87.4	-	[16]
Rice husk	Palm wax	92.1	-	[17]
Rice husk	Soy wax	83.9	-	[17]
Walnut shell	Methyl palmitate	108.3	89.4	[11]
Almond shell	PEG	82.7	60.1	[8]
Pinecone	Palmitic acid	84.7	43.7	[10]
Wood	1-tetradecanol	165.8	114.0	[18]
Activated walnut shell	methyl palmitate (MP)	138.1	57.8	[11]
Sunflower straw	Stearic acid	-	106.2	[19]

In comparison to pure phase transition materials, higher heat transfer rates were obtained, proving the high thermal conductivities. More recently, a study on doping multiwalled carbon nanotubes into phase transition materials for effective thermal energy storage was published by Atinafu *et al.* [20]. It was discovered that composites made of carbon nanotubes and biochar made from bamboo biomass have positively improved the thermal stability and energy-storage capabilities of phase change materials [20]. So, using nanomaterials such as graphene nanosheets and boron nitride nanotubes and sheets, among others, biochar can be effectively prepared using biomass as a cost-effective precursor [21 - 24]. These nanomaterial-engineered biochars made from biomass are promising candidates for producing leak-free composite phase change materials that can be used as functional materials. These materials may be impregnated to create energy-efficient construction building blocks like concrete and plaster for controlling the temperature of buildings.

CONCLUSION

The use of biochar as a potential reinforcement for the development of bio-polymer nanocomposites for thermal energy storage applications is becoming increasingly popular, with researchers highlighting the importance of considering features such as surface morphology, pore size, and crystallinity when producing form-stable composite phase change materials. Biochar made from various biomass sources, including almond shells and walnut shells, has been identified as a viable filler material for modifying the properties of composites. However, it is important to choose the appropriate synthesis technique of biochar from viable raw materials to increase composite properties. Despite this, almond shells and walnut shells have great potential for creating sustainable and eco-friendly bio-polymer nanocomposites for thermal energy storage applications. The next step in this research area is to identify the industrial and commercial applications for biomass-based composites to ensure their effective proliferation across various sectors, promoting the development of a circular bioeconomy and sustainability.

REFERENCES

[1]   Almonds, production quantity (tons) for Germany – Tilasto. Available from: https://www.tilasto.com/en/topic/geography-and-agriculture/crop/almonds/almonds-production-quantity (accessed Oct. 11, 2022).

[2]   Moldero, D.; López-Bernal, Á.; Testi, L.; Lorite, I.J.; Fereres, E.; Orgaz, F. Long-term almond yield response to deficit irrigation. *Irrig. Sci.,* **2021,** *39*(4), 409-420.
[http://dx.doi.org/10.1007/s00271-021-00720-8]

[3]   Remón, J.; Latorre-Viu, J.; Matharu, A.S.; Pinilla, J.L.; Suelves, I. Analysis and optimisation of a novel 'almond-refinery'concept: Simultaneous production of biofuels and value-added chemicals by hydrothermal treatment of almond hulls. *Science of the Total Environment,* **2022,** *765,* 142671. https://www.sciencedirect.com/science/article/pii/S0048969720362008?casa_token=LmWyYlmpzKIAAAAA:dFjpZerAAIib6p3KrfU4huYW9jITPnbvsgwUHk7Hl38lzcYQiTvhxdJpT-XMtjaoAuEmtOGw

[4]   Esfahlan, A.; Jamei, R. The importance of almond (*Prunus amygdalus* L.) and its by-products,Elsevier, **2010,** *120*(2), 349-360. https://www.sciencedirect.com/science/article/pii/S0308814609011236?casa_token=TDXBiExMIDMAAAAA:1xqYHgn18aiAo4cy1M5Sm0CPNZtgB8cg9PgZUdCXPne_bvUcZGkZf1PMm5uBCgODp4yguiHC
[http://dx.doi.org/10.1016/j.foodchem.2009.09.063]

[5]   Guo, T.; Tian, W. Effect of Pore Structure on CO_2 Adsorption Performance for $ZnC_{12}/FeC_{13}/H_2O$(g) Co-Activated Walnut Shell-Based Biochar, Atmosphere **2022,** *13*(7), 110.
[http://dx.doi.org/10.3390/atmos13071110]

[6]   Comas, J. F. Shell hardness in almond: Cracking load and kernel percentage, *Sci Hortic,,* **2019,** *245,* 7-11.
[http://dx.doi.org/10.1016/j.scienta.2018.09.075]

[7]   Demirbaş, A. Estimating of structural composition of wood and non-wood biomass samples. *Energy Sources,* **2005,** *27*(8), 761-767.
[http://dx.doi.org/10.1080/00908310490450971]

[8]   Chen, Y.; Cui, Z.; Ding, H.; Wan, Y.; Tang, Z.; Gao, J. Cost-effective biochar produced from agricultural residues and its application for preparation of high performance form-stable phase change material via simple method. *Int. J. Mol. Sci.,* **2018,** *19*(10), 3055.
[http://dx.doi.org/10.3390/ijms19103055] [PMID: 30301253]

[9]   Sparavigna, A.C. Biochar for Shape Stabilized Phase-Change Materials *ChemRxiv. Cambridge: Cambridge Open Engage,* **2022.**
[http://dx.doi.org/10.26434/chemrxiv-2022-4nthj]

[10] Wan, Y.; Chen, Y.; Cui, Z.; Ding, H.; Gao, S.; Han, Z.; Gao, J. A promising form-stable phase change material prepared using cost effective pinecone biochar as the matrix of palmitic acid for thermal energy storage. *Sci. Rep.,* **2019**, *9*(1), 11535.
[http://dx.doi.org/10.1038/s41598-019-47877-z] [PMID: 31395898]

[11] Hekimoğlu, G. Walnut shell derived bio-carbon/methyl palmitate as novel composite phase change material with enhanced thermal energy storage properties *J Energy Storage,* **2021**, *35*, 102288.
[http://dx.doi.org/10.1016/j.est.2021.102288]

[12] Atinafu, D. G.; Chang, S. J.; Kim, K.-H.; Kim, S. Tuning surface functionality of standard biochars and the resulting uplift capacity of loading/energy storage for organic phase change materials *J. Chem. Eng.,* **2020**, *394*, 125049.
[http://dx.doi.org/10.1016/j.cej.2020.125049]

[13] Zhang, W. Lauric-stearic acid eutectic mixture/carbonized biomass waste corn cob composite phase change materials: Preparation and thermal characterization In: *Thermochim Acta*; , **2019**; 674, pp. 21-27.
[http://dx.doi.org/10.1016/j.tca.2019.01.022]

[14] Gu, X.; Liu, P.; Liu, C.; Peng, L.; He, H. A novel form-stable phase change material of palmitic acid-carbonized pepper straw for thermal energy storage *Mater Lett,* **2019**, *248*, 12-15.
[http://dx.doi.org/10.1016/j.matlet.2019.03.130]

[15] Feng, L.; Zheng, J.; Yang, H.; Guo, Y.; Li, W.; Li, X. Preparation and characterization of polyethylene glycol/active carbon composites as shape-stabilized phase change materials In: *Sol. Energy Mater. Sol. Cells*; **2011**; *95*(2), pp. 644-650.
[http://dx.doi.org/10.1016/j.solmat.2010.09.033]

[16] Nicholas, A.; Hussein, M.; Zainal, Z.; Khadiran, T. Palm kernel shell activated carbon as an inorganic framework for shape-stabilized phase change material. *Nanomaterials (Basel),* **2018**, *8*(9), 689.
[http://dx.doi.org/10.3390/nano8090689] [PMID: 30189654]

[17] Jeon, J.; Park, J.H.; Wi, S.; Yang, S.; Ok, Y.S.; Kim, S. Latent heat storage biocomposites of phase change material-biochar as feasible eco-friendly building materials In: *Environ Res*; , **2019**; 172, pp. 637-648.
[http://dx.doi.org/10.1016/j.envres.2019.01.058]

[18] Yang, H. Low-cost, three-dimension, high thermal conductivity, carbonized wood-based composite phase change materials for thermal energy storage *Energy,* **2018**, *159*, 929-936.
[http://dx.doi.org/10.1016/j.energy.2018.06.207]

[19] Wen, R.; Zhang, W.; Lv, Z.; Huang, Z.; Gao, W. A novel composite Phase change material of Stearic Acid/Carbonized sunflower straw for thermal energy storage *Mater Lett,* **2018**, *215*, pp. 42-45.
[http://dx.doi.org/10.1016/j.matlet.2017.12.008]

[20] Atinafu, D. G.; Wi, S.; Yun, B. Y.; Kim, S. Engineering biochar with multiwalled carbon nanotube for efficient phase change material encapsulation and thermal energy storage *Energy,* **2021**, *216*, 119294.
[http://dx.doi.org/10.1016/j.energy.2020.119294]

[21] Singh, P.; Sharma, R. K.; Khalid, M.; Goyal, R.; Sarı, A.; v Tyagi, V. Evaluation of carbon based-supporting materials for developing form-stable organic phase change materials for thermal energy storage: A review *Sol. Energy Mater. Sol. Cells,* **2022**, *246*, 111896.
[http://dx.doi.org/10.1016/j.solmat.2022.111896]

[22] Kode, V.R.; Thompson, M.E.; McDonald, C.; Weicherding, J.; Dobrila, T.D.; Fodor, P.S.; Wirth, C.L.; Ao, G. Purification and Assembly of DNA-Stabilized boron nitride nanotubes into aligned Films. *ACS Appl. Nano Mater.,* **2019**, *2*(4), 2099-2105.
[http://dx.doi.org/10.1021/acsanm.9b00088]

[23] Kode, V.R.; Hinkle, K.R.; Ao, G. Interaction of DNA-Complexed Boron Nitride Nanotubes and Cosolvents Impacts Dispersion and Length Characteristics. *Langmuir,* **2021**, *37*(37), 10934-10944.

[http://dx.doi.org/10.1021/acs.langmuir.1c01309] [PMID: 34496213]

[24] Yang, J. Largely enhanced thermal conductivity of poly (ethylene glycol)/boron nitride composite phase change materials for solar-thermal-electric energy conversion and storage with very low content of graphene nanoplatelets *J. Chem. Eng.,* **2017**, *315*, 481-490.
[http://dx.doi.org/10.1016/j.cej.2017.01.045]

CHAPTER 5

The Application of Biocarbon Polymer Nanocomposites as Filaments in the FDM Process – A Short Review

Singaravel Balasubramaniyan[1,*], **Niranjan Thiruchinapalli**[2] and **Rutika Umesh Kankrej**[1]

[1] *Department of Mechanical Engineering, Vignan Institute of Technology and Science, Hyderabad, Telangana 508284, India*

[2] *Department of Mechanical Engineering, Mahatma Gandhi Institute of Technology, Hyderabad, Telangana 500075, India*

Abstract: Fused Deposition Modeling (FDM) is a solid-based 3D printing process. It is one of the additive manufacturing technologies that is used to create a three-dimensional (3D) object using a CAD model. In the FDM process, raw material also known as filament, is initially in the solid state. Nowadays, biocarbon-incorporated polymer-based nanocomposite is used as a filament in the FDM process, due to the enhanced strength of the base polymer. In this paper, a review of carbon extracted from natural waste, such as tea powder, coffee grounds, egg shells, ocean plastic, coconut shells, *etc.*, is presented The extraction procedure of biocarbon is given in detail. The results indicate that the strength enhancement of polymers can be achieved by incorporation of derived carbon from industry as well as agriculture waste. In addition, biocarbon-based polymer nanocomposite filaments in the FDM process can be developed by reinforcing the polymer matrix with carbon nanoparticles. Future work of this review process will explore the biobased carbon from various waste resources. The application of biocarbon-based polymer nanocomposites for the 3D printing process is highlighted.

Keywords: 3D Printing Process, Biocarbon Polymer Nanocomposite, FDM, Filament.

1. INTRODUCTION

The 3D printing process is one of the additive manufacturing technologies. It uses a CAD model and introduces it layer by layer to form a three-dimensional object. A CAD model is sliced into layers, generally called Standard Triangular Language (STL). It is classified into the liquid form, solid form, and powder form.

* **Corresponding author Singaravel Balasubramaniyan:** Department of Mechanical Engineering, Vignan Institute of Technology and Science, Hyderabad, Telangana, 508284, India; E-mail: singnitt@gmail.com

Deepa Kodali & Vijaya Rangari (Eds.)

FDM is one of the solid-based 3D printing processes. FDM uses a filament, which is in a solid form. In FDM, the filament is heated in the nozzle, converted into a semi-liquid form, and finally squeezed out. This semi-liquid is deposited onto the platform of the machine and solidified. FDM utilizes a nozzle and moves in different directions, such as x, y, and z, and can be controlled numerically [1, 2].

The filament used in the FDM process is a thermoplastic feedstock. FDM filament involves feeding a spool of filament to the nozzle, in which filament gets heated, melted and forms a required shape on a platform through a controlled direction. Filaments are available in different sizes and colors. Each filament has significant advantages, properties, and limitations. Tensile strength, impact strength, and hardness are the important properties to be considered. Printing speed, adhesion, melting temperature, and ease of printing are some of the important process parameters in the 3D printing process. The selection of appropriate filament plays a significant role in the functional activities of printed specimens [3 - 5].

The use of polymer-based nanocomposites in 3D printing has recently garnered a great deal of interest. In polymer-based nanocomposites, synthetic and bio-based fillers are reinforced to enhance the strength of the base polymer [6, 7]. Bio-based fillers are obtained from various industrial as well as agricultural wastes. Some researchers have attempted to produce carbon from agricultural waste (spent coffee grounds, PET composite, waste coconut shells, ocean plastic, chicken feathers, and starch-based biochar) and industrial waste (from the packaging industry).

2. LITERATURE REVIEW

Mohammed *et al.* [8] derived carbon from chicken feathers and used it for 3D printing. In their work, synthetization was carried out to derive carbon from waste chicken feathers with the help of the pyrolysis process. Potassium hydroxide was used to activate the synthesized carbon. The results of the experimentation revealed that the derived carbon was successfully used in the 3D printing application of electrode preparation of a supercapacitor device. Idrees, *et al.* [9] developed a supercapacitor using the 3D printing process. In their investigation, an electrode was 3D printed with incorporated activated carbon which was synthesized from packaging waste. The results demonstrated that the 3D-printed electrode, with the inclusion of activated carbon, exhibited the ability to withstand high loads. Synthesized carbon from packaging industry waste was used to develop 3D-printed electrodes, thereby leading to enhanced sustainability. Mohammed *et al.* [10] investigated effective reinforcement fillers in polymer nanocomposites. They used carbon derived from starch-based packaging waste as a filler in polymer composites. High-quality biochar was synthesized from packaging waste using the pyrolysis process. The results revealed an improvement in surface area with fewer defects compared to the biochar obtained from natural biomass. 3D printed samples were subjected to mechanical tests and the experimental results showed enhanced strength. Shao *et al.* [11] developed a conductive structure using the 3D printing process. Pyrolysis was performed using lignocellulosic materials. The results indicated that carbon was characterized effectively and thus contributed to improved mechanical and electrical properties of the nanocomposites. Mohammed *et al.* [12] suggested that achieving uniform dispersion and better adhesion were crucial factors in the fabrication of polymer nanocomposites. In their study, surface property modification using low-temperature plasma treatment was performed on biocarbon, which was derived from starch-based packaging waste. The results showed that plasma-based surface treatment has effectively influenced the carbon surface and improved its mechanical strength. Idrees *et al.* [13] developed a polymer composite using consumer-grade Polyethylene terephthalate (PET) and packaging waste derived biocarbon for 3D printing applications. Pyrolysis was used to derive biocarbon from packaging waste. An experimental investigation of surface morphology, textural analysis, and thermal stability of 3D printed specimens was conducted. The results showed improved mechanical properties with the incorporated biocarbon-based polymer nanocomposite. A sustainability approach was attempted utilizing consumer waste for the 3D printing process. Alhelal *et al.* [14] used pyrolysis to derive semi-crystalline biochar from spent coffee grounds. This derived carbon was used as reinforcement in epoxy resin with a ratio of 1 wt. % to 3 wt. % for the direct wire 3D printing process. A mechanical performance analysis was carried out on the 3D printed samples, and the results indicated increased strength with 1 wt %. Also, the strength of 3 wt% reinforced specimen was less at higher loading due to agglomerations. Guptha *et al.* [15] investigated biocomposites developed from recycled consumer PET bottles and biocarbon derived from spent coffee grounds. The pyrolysis process was performed to extract the biocarbon from the spent coffee grounds. The results revealed the enhanced performance of developed biocarbon-based polymer nanocomposites due to better interfacial bonding and improved flexural strength. Thus the developed composite was successfully used in the filament preparation of the 3D printing process. García *et al.* [16] used a 3D printing process to develop a prototype using recycled ocean plastic material-based polymer composite. In their work, the issue of warpage related to the printing of recycled plastic material was discussed. One solution to mitigate this problem was to incorporate fillers into a reinforced blend. Pyrolyzed biocarbon was blended with ocean waste plastic, and filaments were made for 3D printing. Taguchi analyses were performed to execute the experiments. The results of the experiments showed improvement in the mechanical properties. Also, it was noticed that the products developed for 3D printing were more sustainable and impacted the manufacturing economy. Two different prototypes were made, namely a gear and frame, using the 3D printing process. Diederichs *et al.* [17] developed a novel 3D printing raw material using poly (trimethylene terephthalate) (PTT) and pyrolyzed Miscanthus biocarbon. In their work, 5% and 10% wt blending ratios were used to develop 3D-printed specimens. Mechanical performance analysis was carried out and the results revealed that 5 wt% 3D printed specimens performed better compared to other specimens. Mayakrishnan *et al.* [18] used the pyrolysis process to extract biocarbon from a digitalis purpurea plant mixed with thermoplastic polyurethane (TPU) for filament development for the 3D printing process. The mechanical performance of the polymer-based nanocomposite was investigated by varying the concentrations of biocarbon. The findings indicated that varying concentrations of biocarbon additives had a notable impact on the mechanical performance. Kodali *et al.* [19] used coconut shell powder for the extraction of carbon and fabricated polycarbonate filament for the 3D printing process. The effect of different weight concentrations of the loaded fillers on the filament was investigated. Mechanical tests were conducted, and the results indicated that the carbon-incorporated polycarbonate filament performed better than the neat filament. Umerah *et al.* [20] prepared filaments for the 3D printing process using waste coconut shell-derived carbon incorporated polymer composite. The effect of different ratios of carbon nanoparticles on the mechanical strength was investigated. The strength of 3D printed specimens was improved with optimum levels of nanoparticle ratio. Miswan *et al.* [21] used carbon derived from coconut husks for polymer composites to perform 3D printing. The results revealed that 3D-printed agriculture waste-based polymer composites can be used in different applications. Rangel-Sequeda *et al.* [22] conducted 3D printing using activated carbon incorporated into carboxymethylcellulose to investigate mechanical strength. Direct ink writing was used to obtain a monolithic form of activated carbon powder. The results revealed that the 3D printing of carbon with carboxymethylcellulose as a binder has impressive compressive strengths ranging from 0.58 to 3.53 Mpa. Radhika *et al.* [23] reviewed the various applications of polymer nanocomposites incorporated with biocarbon in the FDM-based 3D printing process. Their review process was focused on biomaterial applications in the 3D printing process. Furthermore, this review delved into the acquisition of sustainable products through the 3D printing process. The effect of the filler on the filament was focused in detail and summarized. The review also summarized that polymer nanocomposites with biocarbon can be developed using two distinct approaches: reinforcing a biobased polymer with nanoparticle fillers and incorporating a biobased material as filler within the polymer. From the literature, it is noticed that the derived bio-carbon from various industrial wastes as well as agricultural wastes can be incorporated into polymers for developing 3D printable filaments. Table **1** summarizes the studies investigating the utilization of biocarbon derived from various waste resources for the development of polymer nanocomposites.

Table 1. Previous studies on biocarbon polymer nanocomposite.

S. No	Reference Number and Year	Biocarbon Derived from Industrial or Agricultural Waste
1	Mohammed *et al.* [8]	chicken feathers
2	Idrees *et al.* [9]	packaging waste
3	Mohammed *et al.* [10]	packaging waste
4	Shao *et al.* [11]	cellulose/lignosulfonate/ cellulose powder
5	Mohammed *et al.* [12]	starch-based packaging waste
6	Idrees *et al.* [13]	Consumer used PET bottles
7	Alhelal *et al.* [14]	spent coffee grounds
8	Guptha *et al.* [15]	spent coffee grounds
9	García *et al.* [16]	ocean plastic material
10	Diederichs *et al.* [17]	Miscanthus biocarbon
11	Mayakrishnan *et al.* [18]	Digitalis purpurea plants
12	Kodali *et al.* [19]	coconut shell powder
13	Umerah *et al.* [20]	coconut shell powder
14	Miswan *et al.* [21]	coconut husks
15	Rangel-Sequeda *et al.* [22]	Monolithic activated carbon obtained from direct ink writing

3. DISCUSSION

The pyrolysis process is used to synthesize and extract carbon from chicken feather waste. Initially, chicken feather waste was cleaned with water to remove dirt and blood. Some steps included drying with air, ball milling, and a two-step heat treatment in a furnace. Researchers have estimated that the surface area and pore volume of activated carbon were 515.896 m^2 and 0.607 cc/g, respectively. The 3D printing process was used to develop the filament with the help of synthesized activated carbon [8]. The following steps (Table **2**) were used for carbon synthesis using chicken feather waste [8]. The results of X-ray diffraction indicated that a high amount of amorphous porous carbon was present. The results of Raman's investigation indicated an increase in the amorphous nature of carbon. This synthesized carbon incorporated in the polymer was used to print to an electrode, which found its application in supercapacitors successfully. The capacitance value of the 3D-printed electrode was 310.32 mF/cm^2. The authors also observed improvement through nanoindentation results.

Table 2. Carbon synthesis using chicken feather waste.

Step 1	Collection and cleaning of feathers using detergent and distilled water
Step 2	The ball milling process (3 Hours)
Step 3	Sieved to 45 microns
Step 4	High pressure and temperature pyrolysis
Step 5	Ground to form uniform size.

Idrees *et al.* [9] used starch from packaging waste to derive carbon using the pyrolysis process. Further, ultrasonication was applied to biocarbon to reduce particle size and increase surface area. Polypropylene polymer was incorporated with this biocarbon with different loadings ranging from 0 to 1 wt %. Owing to the enhanced surface area and size reduction, the proposed biocarbon was used as an effective reinforcement in the polymer nanocomposite [9]. Ultrasonication of the biocarbon was used to enhance the reinforcing ability. Filaments for the FDM process were developed, and 3D-printed tensile specimens were prepared. Table **3** shows the steps involved in carbon extraction using packaging waste and the applications of 3D printing. XRD and Raman studies revealed that more graphitic carbon was present. The tensile strength and modulus were observed to be high. Biochar was incorporated into the polymer at varying concentrations ranging from 0 to 1 wt%. The findings demonstrated a 34% enhancement in tensile strength and a 46% increase in modulus strength with the inclusion of 0.75 wt%. The results also indicated that using biochar-filled filament facilitated smooth printing without any printer head clogs.

Table 3. Steps involved in carbon extraction, filament making, and the 3D printing process.

Step 1	Manual and mechanical mixing of biochar and polypropylene polymer.
Step 2	The extrusion process
Step 3	Pelletizing and filament extrusion for the uniform diameter of the filament.
Step 4	3D printing as per the ASTM standards.
Step 5	Tensile test using an ultimate tensile testing machine.

Another study reported the efficient use of biocarbon derived from consumer-used recycled PET bottles to make biocarbon/PET composite filament for 3D Printing. Mechanical performance testing was conducted and the results revealed an improvement in tensile strength. Biocarbon-incorporated 3D printed parts performed better in terms of mechanical strength due to improved bonding properties. Moreover, this study showed other advantages, such as dimensional stability, low value of thermal stress, and dimensional accuracy [13]. Table **4** shows the steps involved in the development of composite filament making for 3D printing. The investigation's findings suggested that incorporating 0.5 wt% biochar into PET led to a notable 32% increase in tensile strength. This improved tensile strength suggested the enhanced bonding of biochar and PET. Furthermore, it was noticed that PET nucleation and crystallinity improved due to the reinforcement of biochar.

Table 4. Fabrication of composite filament for 3D printing.

Step 1	The selection of PET and biochar
Step 2	Filament extrusion
Step 3	The 3D printing process

Alhelal *et al.* [14] utilized spent coffee grounds (SCG), subjecting them to a high-pressure/temperature reactor for pyrolysis, and subsequently employed ultrasonication to obtain biochar. This biochar was directly added to the epoxy and hardener. The stirring process was carried out frequently. This composite was used to produce a 3D printing filament. The direct wire method was used for the fabrication of filament with the addition of 1 to 3 wt % of biochar. The inclusion of biochar proved beneficial in rendering the resin more viscous in nature, thereby preventing resin clogging at the printer head. The specimen with 1 wt% of biochar improved the tensile strength by 43.30%. The addition of biochar up to 3% drastically reduced the tensile strength. Table **5** presents the various procedures adopted for composite filament fabrication for the 3D printing process.

Table 5. 3D printing using composite filaments from spent coffee grounds.

Step 1	The selection of epoxy and biochar
Step 2	Ultrasonication
Step 3	Addition of the hardener
Step 4	The 3D printing process
Step 5	Testing process

In the FDM process, warpage during printing is predominant, especially with the use of recycled plastic. Polymer reinforced with filler can increase the strength of the base polymer. The combination of recycled ocean plastic and industrial waste of agro products has provided sustainable products for 3D printing applications. It also leads to warpage-free products during the 3D printing process [16]. Table **6** outlines the different steps comprising the carbon extraction process when combining polymer with recycled ocean plastic for 3D printing applications. In their work, Maldonado-García *et al.* used the pyrolysis process on ocean plastic waste. The biocarbon infused polymer was noted to have 15% improvement in the Young's modulus compared to injection molded parts. Utilizing recycled plastics collected from ocean shores, along with the incorporation of biocarbon, offers a combined advantage for producing products that are resistant to warping. A metal flow index study was also carried out by the authors, and the results suggested that there was no triggering over extrusion on the print head.

Table 6. Carbon extraction from recycled ocean plastic.

Step 1	Pyrolysis and the ball milling process
Step 2	The extrusion process
Step 3	Filament fabrication
Step 4	The 3D printing process

Umerah *et al.* [20] incorporated carbon nanoparticles into PLA, a biodegradable plastic. These carbon nanoparticles were synthesized using the pyrolysis procedure and were derived from waste coconut shells. Carbon nanoparticles were blended with different ratios (0.2 wt %, 0.6 wt %, and 1wt %) and were made into filaments for the 3D printing process. Mechanical strength was varied with different weight ratios and the results indicated that 0.6 wt % specimens showed better mechanical strength compared to the other specimens. Further addition of carbon nanoparticles (1wt %) to the polymer matrix led to an agglomeration effect and reduced strength [20]. The characterization was carried out using XRD and Raman analysis to show the various properties of carbon nanoparticles. These XRD and Raman analysis were used to understand the nature of carbon and its degree of graphitization. The results indicated that the Raman peak of ball-milled coconut shell powder was at 1323.7 cm^{-1} suggesting the graphitic nature of carbon. Surface morphology analysis revealed that the carbon powder has an irregular shape after the ball milling process. Mechanical strength exhibited enhancement at nanoparticle weight percentages of 0.2, 0.6, and 1 wt% incorporated into the polymer. The tensile strength experienced a 50% increase as a result of adding nanocarbon up to 0.6 wt% into the base polymer.

CONCLUSION

The following conclusions were obtained from the review:

1. It is understood that the polymer-based nanocomposite reinforced with biocarbon as a filler material can be used to enhance the strength of the base polymer.

2. Biocarbon can be successfully derived from various agricultural wastes (spent coffee grounds, PET composite, waste coconut shells, ocean plastic, chicken feathers, and starch-based biochar) and industrial waste (waste from the packaging industry).

3. Reinforcement of bio-based carbon nanoparticles as fillers in polymer filaments significantly improves the mechanical properties.

4. Further investigation is required to assess the performance of biocarbon polymer nanocomposite materials in relation to FDM process parameters.

5. Various concentrations of the derived carbon added to the base polymer were investigated. The results suggested that high weight percentages lead to the agglomeration of added biocarbon.

6. Polymer-based nanocomposites are employed as filaments in the 3D printing process, and they often encounter issues related to warping during printing. It has been observed that the incorporation of derived biocarbon as a filler significantly impacts the strength of the base polymer.

7. High load capacity is noticed with biocarbon incorporated polymer nanocomposite, due to its better surface morphology, bonding, and adhesion properties.

8. Biocarbon and its polymer nanocomposite are used to enhance sustainability. Most of the biocarbon extraction uses industrial and agricultural waste. The pyrolysis process is commonly employed for extracting biocarbon from industrial and agricultural waste.

9. Reinforcing capability is enhanced for the derived biocarbon through an ultrasonication process performed after pyrolysis. This is used to increase the overall capability of the composite prepared for 3D printing.

10. Future research should focus on conducting optimization studies regarding the incorporation of biocarbon, specifically with regard to enhancing mechanical strength.

REFERENCES

[1]   Rashed, K.; Kafi, A.; Simons, R.; Bateman, S. Fused filament fabrication of nylon 6/66 copolymer: Parametric study comparing full factorial and Taguchi design of experiments. *Rapid Prototyp. J,* **2022,** 1111-1128.
[http://dx.doi.org/10.1108/RPJ-06-2021-0139]

[2]   Heidari-Rarani, M.; Ezati, N.; Sadeghi, P.; Badrossamay, M. R. Optimization of FDM process parameters for tensile properties of polylactic acid specimens using Taguchi design of experiment method. *J. Thermoplast. Compos. Mater,* **2020,** 0892705720964560.

[3]   Kumar, R.; Sharma, H.; Saran, C.; Tripathy, T.S.; Sangwan, K.S.; Herrmann, C. A comparative study on the life cycle assessment of a 3D printed product with PLA, ABS & PETG materials. *Procedia CIRP,* **2022,** *107,* 15-20.
[http://dx.doi.org/10.1016/j.procir.2022.04.003]

[4]   Atakok, G.; Kam, M.; Koc, H.B. A Review of mechanical and thermal properties of products printed with recycled filaments for use in 3d printers. *Surf. Rev. Lett.,* **2022,** *29*(2), 2230002.
[http://dx.doi.org/10.1142/S0218625X22300027]

[5]   Kichloo, A. F.; Aziz, R.; Haq, M. I. U.; Raina, A. Mechanical and physical behaviour of 3D printed polymer nanocomposites-a review. *J. Industrial and Sys. Eng.,* **2021,** *38*(4), 484-502.
[http://dx.doi.org/10.1504/IJISE.2021.116929]

[6]   Madhu, N.R.; Erfani, H.; Jadoun, S.; Amir, M.; Thiagarajan, Y.; Chauhan, N.P.S. Fused deposition modelling approach using 3D printing and recycled industrial materials for a sustainable environment: A review. *Int. J. Adv. Manuf. Technol.,* **2022,** *122*(5-6), 2125-2138.
[http://dx.doi.org/10.1007/s00170-022-10048-y] [PMID: 36091410]

[7]   Andrzejewski, J.; Gronikowski, M.; Aniśko, J. A novel manufacturing concept of LCP fiber-reinforced gpet-based sandwich structures with an FDM 3D-printed core. *Materials,* **2022,** *15*(15), 5405.
[http://dx.doi.org/10.3390/ma15155405] [PMID: 35955339]

[8]   Mohammed, Z.; Jeelani, S.; Korivi, N. S.; Rangari, V. *Synthesis and characterization of N-doped porous carbon from chicken feathers for 3D printed electrode applications.,* **2022.**
[http://dx.doi.org/10.21203/rs.3.rs-2230929/v1]

[9]   Idrees, M.; Ahmed, S.; Mohammed, Z.; Korivi, N.S.; Rangari, V. 3D printed supercapacitor using porous carbon derived from packaging waste. *Addit. Manuf.,* **2020,** *36,* 101525.
[http://dx.doi.org/10.1016/j.addma.2020.101525]

[10]   Mohammed, Z.; Jeelani, S.; Rangari, V. Effective reinforcement of engineered sustainable biochar carbon for 3D printed polypropylene biocomposites. *Composites Part C: Open Access,* **2022,** *7,* 100221.
[http://dx.doi.org/10.1016/j.jcomc.2021.100221]

[11]   Shao, Y.; Guizani, C.; Grosseau, P.; Chaussy, D.; Beneventi, D. Use of lignocellulosic materials and 3D printing for the development of structured monolithic carbon materials. *Composites Part B: Engineering,* **2018,** *149,* 206-215.
[http://dx.doi.org/10.1016/j.compositesb.2018.05.035]

[12]   Mohammed, Z.; Jeelani, S.; Rangari, V.K. Effect of low-temperature plasma treatment on starch-based biochar and its reinforcement for three-dimensional printed polypropylene biocomposites. *ACS Omega,* **2022,** *7*(44), 39636-39647.
[http://dx.doi.org/10.1021/acsomega.2c02372] [PMID: 36385856]

[13]   Idrees, M.; Jeelani, S.; Rangari, V. Three-dimensional-printed sustainable biochar-recycled PET composites. *ACS Sustain. Chem.& Eng.,* **2018,** *6*(11), 13940-13948.
[http://dx.doi.org/10.1021/acssuschemeng.8b02283]

[14]   Alhelal, A.; Mohammed, Z.; Jeelani, S.; Rangari, V.K. 3D printing of spent coffee ground derived biochar reinforced epoxy composites. *J. Compos. Mater.,* **2021,** *55*(25), 3651-3660.

[http://dx.doi.org/10.1177/00219983211002237]

[15] Gupta, A; Mohanty, A. K.; Misra, M. Biocarbon from spent coffee ground and their sustainable biocomposites with recycled water bottle and bale wrap: A new life for waste plastics and waste food residues for industrial uses. *Compos Part A-Appl,* **2022**, *154*, 106759.
[http://dx.doi.org/10.1016/j.compositesa.2021.106759]

[16] Maldonado-García, B.; Pal, A.K.; Misra, M.; Gregori, S.; Mohanty, A.K. *Sustainable 3D printed composites from recycled ocean plastics and pyrolyzed soy-hulls: Optimization of printing parameters, performance studies and prototypes development., Compos. C: Open Access,* **2021**, 100197.

[17] Diederichs, E.; Picard, M.; Chang, B.P.; Misra, M.; Mohanty, A. Extrusion based 3D printing of sustainable biocomposites from biocarbon and poly *(trimethylene terephthalate). Molecules,* **2021**, *26*(14), 4164.
[http://dx.doi.org/10.3390/molecules26144164] [PMID: 34299439]

[18] Mayakrishnan, V.; Mohamed, J.K.; Selvaraj, N.; SenthilKumar, D.; Annadurai, S. Effect of nano-biochar on mechanical, barrier and mulching properties of 3D printed thermoplastic polyurethane film. *Polym. Bull.,* **2023**, *80*(6), 6725-6747.
[http://dx.doi.org/10.1007/s00289-022-04380-2]

[19] Kodali, D.; Umerah, C.O.; Jeelani, S.; Rangari, V. K. Fabrication of polycarbonate filaments infused with carbon from coconut shell powder for 3D printing applications. *In Novel Applications of Carbon Based Nano-materials,* **2022**, 225-234.
[http://dx.doi.org/10.1201/9781003183549-14]

[20] Umerah, C.O.; Kodali, D.; Head, S.; Jeelani, S.; Rangari, V.K. Synthesis of carbon from waste coconutshell and their application as filler in bioplast polymer filaments for 3D printing. *Compos.B: Eng.,* **2020**, 108428.

[21] Miswan, R.I.; Mokhtar, M.F.; Idris, F.M. Fabrication and characterization of low-cost raw coconut husk and activated carbon coconut husk/epoxy-3d printed honeycomb structure as an electromagnetic wave. *in Proc. KOSIST,* **2020**.

[22] Rangel-Sequeda, J.F.; Loredo-Cancino, M.; Águeda Maté, V.I.; De Haro-Del Rio, D.A.; Dávila-Guzmán, N.E. 3D printing of powdered activated carbon monoliths: Effect of structuring on physicochemical and mechanical properties and its influence on the adsorption performance. *Mater. Today Commun.,* **2022**, *33*, 104758.
[http://dx.doi.org/10.1016/j.mtcomm.2022.104758]

[23] Mandala, R.; Bannoth, A.P.; Akella, S.; Rangari, V.K.; Kodali, D. A short review on fused deposition modeling 3D printing of bio-based polymer nanocomposites. *J. Appl. Polym. Sci.,* **2022**, *139*(14), 51904.
[http://dx.doi.org/10.1002/app.51904]

CHAPTER 6

Tensile Characteristics of FDM 3D Printed PBAT/PLA/Carbonaceous Biocomposites

Gustavo F. Souza[1], Rene R. Oliveira[2], Janetty J.P. Barros[1,3], Fernando L. Almeida[4] and Esperidiana A.B. Moura[1,*]

[1] *Centro de Química e Meio Ambiente, Instituto de Pesquisas Energéticas e Nucleares, Av. Prof. Lineu Prestes, 2242, São Paulo 05508-000, SP, Brazil*

[2] *Centro de Ciências e Tecnologia de Materiais, Instituto de Pesquisas Energéticas e Nucleares, Av. Prof. Lineu Prestes, 2242, São Paulo 05508-000, SP, Brazil*

[3] *Academic Unit of Materials Engineering, Federal University of Campina Grande, Campina Grande 58249-140, Brazil*

[4] *Faculdade de Tecnologia do Estado de São Paulo "Prof. Miguel Reale", Av. Miguel Ignácio Curi, 360, São Paulo 08295-005, SP, Brazil*

Abstract: The use of carbonaceous fillers in polymeric biocomposite materials has been widely studied due to their potential to add better engineering properties to biocomposites and expand their field of applications. Currently, due to the growing global concerns over environmental pollution and climate change, carbonaceous fillers derived from biomass are the preferred choice for production of the sustainable biocomposite materials. Rice husk ash (RHA), an abundant and sustainable carbonaceous filler obtained from the burn of rice husk in kilns of the processing rice was incorporated into the PBAT/PLA blend. The influence of RHA loading on the tensile properties of FDM-3D printed samples was investigated. Neat PBAT/PLA filament and its biocomposite filaments with 2.5, 5.0 and 7.5 wt. % RHA were prepared by the extrusion process. The filaments were characterized by FTIR, TG, and SEM. FDM-3D printed specimens were subjected to tensile tests.

Keywords: 3D printing, Biocomposites, Carbonaceous Fillers, RHA, SEM, Tensile Properties.

1. INTRODUCTION

The use of carbonaceous fillers in polymeric biocomposite materials has garnered attention due to potential for enhanced engineering properties and their field of

---

* **Corresponding author Esperidiana A.B. Moura:** Centro de Química e Meio Ambiente, Instituto de Pesquisas Energéticas e Nucleares, Av. Prof. Lineu Prestes, 2242, São Paulo 05508-000, SP, Brazil; Tel: 55 11 2810 8320; Fax: 55 11 2810 8320; E-mail: eabmoura@ipen.br

applications. Besides simple processing, lightweight character, cost-effective, and superior mechanical reinforcement, carbonaceous fillers are also important fillers to impart heat dissipation properties, electrical and thermal conductivities [1,2]. Currently, the interest in the conversion of agricultural and forest waste biomass to carbonaceous fillers and their use in polymeric biocomposites has increased notably. The growing global concerns over environmental pollution aspects, climate change, and the necessity to conserve material resources, make these carbonaceous fillers a preferred choice for the production of advanced sustainable biocomposites with enhanced properties. They are considered eco-friendly carbon precursors, which are readily available, economically feasible, green, and renewable in nature. Furthermore, the conversion of biomass into carbonaceous fillers and its use in the production of advanced sustainable biocomposites can also contribute to solid waste management [3, 4].

The FDM 3D printing process, an additive manufacturing (AM) technique, finds applications in various industrial fields. This process enables the production of complex parts by depositing successive layers of thermoplastic material filament. FDM 3D printing processing has been used in many industrial sectors, such as aerospace, automotive, food, and medical. In the medical industry, this technology has been adopted to manufacture orthopedic devices as prostheses as well as other devices, bone tissue engineering, and the development of scaffolds with different porosity. Other segments like home appliances and creative gifts have also benefited from the 3D printing process. The increased interest in this technology is because of its easy use, simple operation, low cost, and facile production of complex geometries [5, 7].

The parameters for fused filament 3D printing are fundamental characteristics for deposition of materials. Some configurations allow for greater physical and mechanical resistance, beyond impacting the visual quality of prints while some parameters are notable for the efficient fluid functioning of 3D printing. The process of slicing 3D parts depends on the technical settings of the equipment [8]. 3D Printers have an extrusion assembly, located at its end, consisting of a temperature-controlled metal nozzle, allowing the material to be continuously melted for layer-by-layer deposition [6]. The nozzle has an orifice from which the polymer is expelled followed by the deposition on the bed. 3D printing machines were generally equipped with 0.4 mm thick nozzles, which is the most common configuration for 3D printing by FDM. Therefore, the printing wall thickness was set according to the original 3D printer nozzle configuration. Other alterations to the slicing process that are unrelated to the machine's physical architecture allow for a greater degree of control to be exercised over the prints that are produced [5]. Some of them are discussed below:

• **Layer height** is the relative measurement of the extrusion between layers. It is the height of each layer that is deposited to form the part. This measure directly influences speed, appearance physique, and endurance. Smaller measures include lower speed, better visual finish, and greater resistance.

• **The Walls** of a 3D object is represented by the part's outer fill. This is responsible for forming the physical and visual identity of the exterior of the piece. The walls influences the strength and quality of the object. Greater number of walls: better resistance and print quality.

• **Flow** is the parameter that defines the percentage of the amount of extrusion to be performed during printing. That is, according to the percentage, the printer deposits a greater or lesser amount of melted filament.

• **The infilling** is responsible for setting in a percentage of the amount of material deposited inside the part, *i.e.*, with 100% fill, parts are completely solid. In the slicing software, it is also possible to modify the filling angles. Commonly, for resistance tests, impressions are made with filling angles of 0°, 45° and 90°.

• **Printing temperatures** always vary as specified by the filament manufacturer. Poor temperature calibration can cause warping, filling, and printing problems. For some polymers, there is a need to use the table heating system, to ensure the adhesion of the material on its surface.

• **Speed** is an important parameter to have a close expectation regarding the time it takes to print any part. Therefore, the speed variation allows for reduction or increasing the printing time. The increase in speed can lead to reduced visual quality and increase the probability of errors during printing. This parameter is given in relation to distance (mm) per time (s).

Conventional thermoplastic materials like polycarbonate (PC), polyamide (PA), acrylonitrile butadiene styrene (ABS), ethylene vinyl acetate (EVA), and the biodegradable polylactic acid (PLA) are usually used to produce FDM 3D printing filaments. However, with the increasing attention to environmental issues, biopolymers such as PLA, and their biocomposites and blends became the preferred choice for FDM 3D printing filaments instead of petroleum-based ones due to their biodegradability and environmental friendliness [8 - 11].

Rice husk (RH) is removed during the industrial process of rice milling and polishing. A large part of this material is used as fuel in rice processing kilns, generating a significant amount of rice husk ash (RHA) as waste. Unfortunately, not all the amount of available rice husk is used as fuel, part of it is still discarded in landfills as waste, contributing to the increase in environmental pollution. RH has about 20% silica and pyrolysis of RH can remove the volatile carbonaceous organic materials present and produce RHA. Therefore, the part of RH currently discarded as waste can be reused for the production of RHA through the pyrolysis process. RHA basically contains silica (around 90%) and carbon, which, added to its low cost and low density, high strength, and modulus, make them excellent materials for application as carbonaceous fillers in polymeric composites. Furthermore, RHA is renewable, sustainable, green, and safe for health [12 - 15].

This work investigates the tensile characteristics of PBAT/PLA/carbonaceous biocomposite specimens that were FDM 3D printed with 100% infill pattern and with varying orientation angles of 0, 45, and 90-degree. In addition, the biocomposites and filaments produced by the melting extrusion process by incorporation of different amounts of the rice husk ash (RHA) into neat PBAT/PLA, are also evaluated by FTIR, TG, and SEM analyses.

2. MATERIALS AND METHODS

2.1. Materials

The main materials used to prepare the PBAT/PLA/RHA biocomposites were: PBAT/PLA Blend (Ecovio® T2308, "BASF SE"), MFI: 9.5 cm^3 / 10 min (ISO 1133), density: 1.4 g cm^{-3} (ISO 1183); and Rice husk ash (RHA), a waste donated by the Brazilian rice industry.

2.2. Rice Husk ash (RHA) Preparation

Rice husk ash (RHA) waste was firstly put in an oven with an air circulation system for 24 h, at 80 ± 2 °C to dry, and then reduced to particle sizes of 125 mm using a ball mill, and re-dried at 80 ± 2°C for 24 h.

2.3. Composite Preparation

The PBAT/PLA blend before processing was placed in an oven with an air circulation system, at 80 °C for 24 hours to keep the moisture present in the blend below 2%. The biocomposites of PBAT/PLA blend with RHA reinforcements, according to composition presented in Table **1**. The blend with RHA was hand premixed and then processed in a Thermo Scientific HAAKE twin-screw co-rotating extruder, model Rheomex 332p, with a 16 mm screw diameter and a length-diameter-ratio of 25. The biocomposites were processed in a two-step extruder to obtain better dispersion of carbonaceous fillers into the PBAT/PLA matrix. Firstly, hand-premixed PBAT/PLA/RHA was melt-mixed with a temperature profile of 150/165/175/175/165/150 °C. The screw speed was set at 30 rpm. The extruded biocomposites (strands) were cooled down, pelletized, and kept in an oven with an air circulation system at 80 °C for 4 hours to dry.

Table 1. PBAT/PLA/RHA biocomposites composition.

Materials	PBAT/PLA (wt. %)	Rice Rusk Ashe (wt. %)
Neat PBAT/PLA	100	0
PBAT/PLA/RHA - 2.5%	97.5	2.5
PBAT/PLA/RHA - 5.0%	95	5.0
PBAT/PLA/RHA - 7.5%	92.5	7.5

2.4. Filament Preparation

PBAT/PLA/RHA biocomposites with different amounts of RHA were fed into Thermo Scientific HAAKE twin-screw co-rotating extruder and the filaments with 1.75±0.01 mm of diameter which are suitable for FDM were produced. The temperature profile was 145/160/175/175/165/145 °C, screw speed set at 30 rpm, and temperature of cooling water of the filaments of 35 °C. The main processing steps of the biocomposites and filaments are schematically presented in Fig. (**1**).

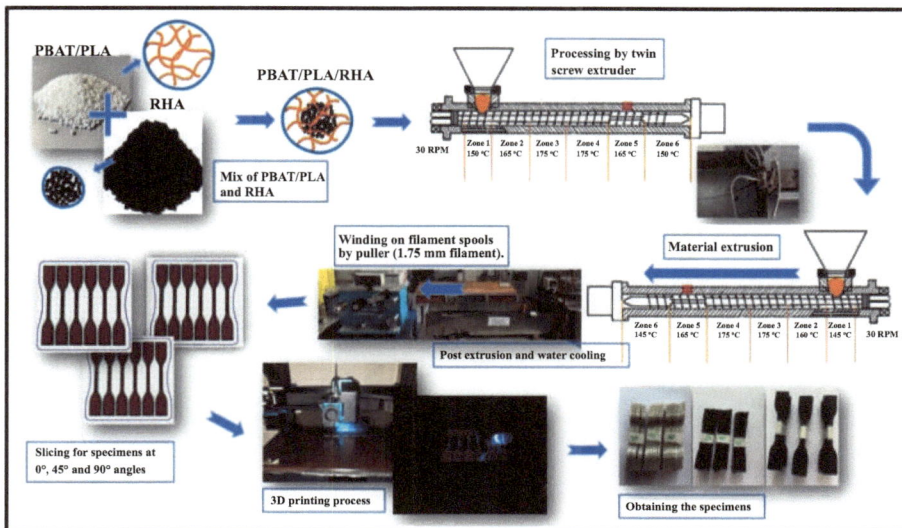

Fig. (1). Schematic of main processing steps of the biocomposites and filaments.

2.5. FDM 3D Printing

The molds for 3D printing were obtained from the digital model supply website Thingiverse [16]. The mold used for the tensile tests complies with the standards of ASTM D638, IV model, [5, 6, 11, 17] as shown in Fig. (2).

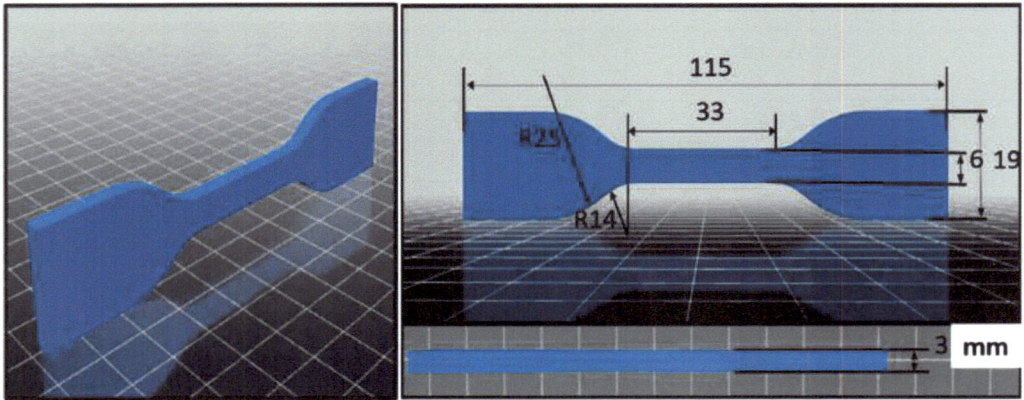

Fig. (2). 3D model from Thingiverse [19].

The samples were made of the same IV mold of the ASTM D638 standard [5, 6, 11, 17]. Printing bed was configured into 6 parts for each 0°, 45° and 90° angle [5, 6] and four composite materials: PLA/PBAT, without RHA; PLA/PBAT/RHA 2.5%; PLA/PBAT/RHA 5%; and PLA/PBAT/RHA 7.5%. Fig. (3) shows the slicing for each angular orientation.

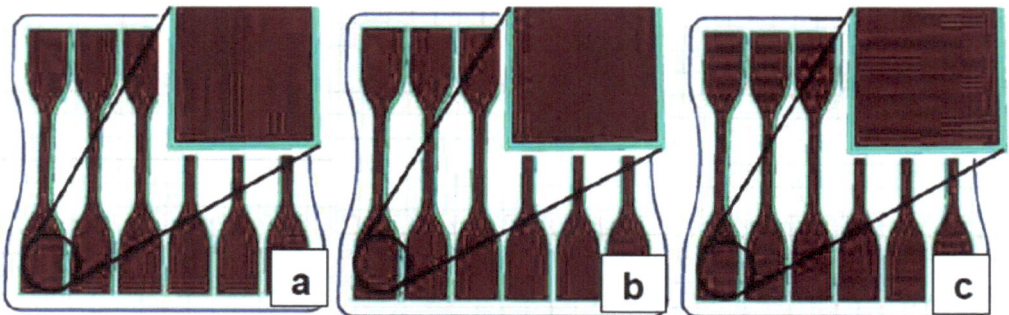

Fig. (3). Slicing for angular orientation of (a) 0°, (b) 45° and (c) 90°.

All models with different materials were printed with the same nozzle temperature of 180 ± 2 °C. The print bed has been set to remain at 70 ± 2 °C. The layer height was set to 0.3 mm. Being the most suitable for the worked materials, the number of walls was set to 2 and speed was set to 50 mm s^{-1}. Flow was defined as 115%, due to the variation in the filament diameter which is observed to be slightly less than 1.75 mm. Table **2** presents the parameters of the 3D printing process of the samples.

Table 2. Parameters of the 3D printing process of the samples.

Hotend (°C)	Bed (°C)	Layer (mm)	Nozzle (mm)	Flow (%)	Speed (mm s^{-1})
180 ± 2	70 ± 2	0.30 ± 0.01	0.40 ± 0.01	115.0 ± 2.5	50.0 ± 0.5

The printed specimens were obtained for the four biocomposite materials with varying orientation angles of 0, 45 and 90-degree, totalizing 24 units for each orientation (72 samples). The specimens were obtained by FDM printer Artillery Sidewinder X1 (Fig. **4**).

Fig. (4). 3D printing of the samples.

2.6. Characterization Methods

2.6.1. Thermogravimetric Analyses (TGA)

TGA analyses were carried out to evaluate the influence of the RHA on the thermal stability of PBAT/PLA/RHA biocomposites, and also the effect of the temperature used during processing by 3D FDM printing on the degradation of the materials. Thermogravimetric analyses using the TGA/SDTA 851e, from

Mettler-Toledo, conducted on samples of about 10 mg. These samples were heated from 25 to 600°C at a heating rate of 20°C/min under a nitrogen atmosphere.

2.6.2. Fourier Transform Infrared Spectroscopy (FTIR)

FTIR analyses were performed in a Perkin Elmer - Frontier in attenuated total reflectance (ATR) mode with 16 scans. The resolution was 4 cm^{-1} and the wavenumber was from 4000 to 650 cm^{-1}.

2.6.3. Scanning Electron Microscopy (SEM)

SEM analyses were conducted in a Model LX-30 Philips instrument operated at an electron acceleration voltage of 12 kV. The samples' cryo-fractured surfaces under liquid nitrogen were sputtered with a gold layer before SEM analyses of the samples' surfaces. Considering that the morphology of the filler-matrix interface plays an important role in the mechanical reinforcement process, the correlation between the morphology of samples' surfaces with different RHA contents and the mechanical performance of the specimens of PLA/PBAT/RHA printed was also evaluated.

2.6.4. Mechanical Testing

Printed specimens from the neat PLA/PBAT and its biocomposites were subjected to tensile tests on Instron model 5564, according to ASTM D 638-99, at room temperature at a cross-head speed of 50 mm min^{-1}. For each composition, five specimens were tested, and the average value of five specimens with the standard deviation was recorded.

3. RESULTS AND DISCUSSION

3.1. TGA

The influence of RHA contents on the biocomposites' thermal stability and the effects of specimens' processing temperature by 3D printing on the degradation of the neat PBAT/PLA and its biocomposites were evaluated by TGA. Fig. (**5**) shows the curves of TGA and DTG (derivative thermogram) of neat PBAT/PLA and its biocomposites. As shown in Fig. (**5**), when the RHA was added to the biocomposites, the onset temperature (T_{onset}) was lower compared to the T_{onset} of neat PBAT/PLA. The T_{onset} represents the temperature at which the weight loss (WL) of materials begins, and T_{max} represents the temperature of the maximum point of thermal degradation of the materials and is the peak of the derived DTGA curves. According to the obtained DTG curves (Fig. **5b**), the neat PBAT/PLA and its biocomposites present two steps of thermal degradation. A similar trend has been observed in the literature [17, 18].

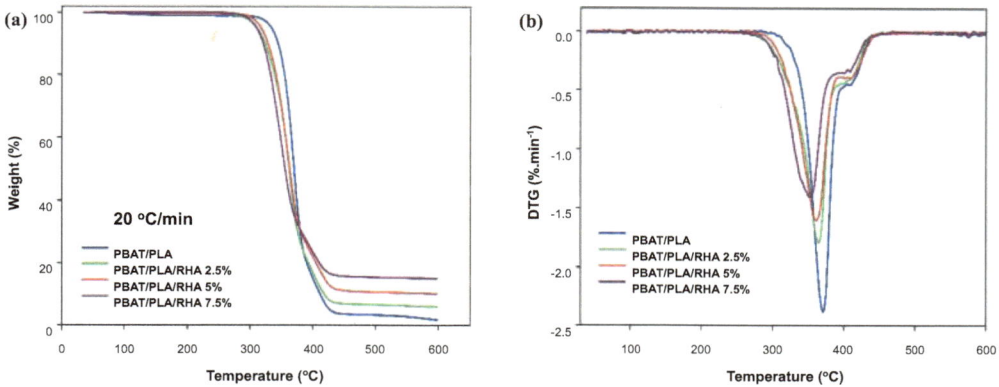

Fig. (5). TGA (**a**) and DTG (**b**) curves of the neat PBAT/PLA and its biocomposites.

The characteristic values determined from TGA for neat PBAT/PLA and its biocomposites are presented in Table **3**. As observed in Table **3**, the incorporation of 2.5 wt. % of RHA results in only insignificant thermal differences at the temperature for 5% weight loss and a decrease of approximately 7°C in temperature for 50% weight loss. Additionally, there is an almost 8°C reduction in the maximum temperature (Tmax) compared to neat PBAT/PLA. According to values presented in Table **3**, the reduction of the temperature for both, 5% and 50% of weight loss was more significant with increasing RHA content. For example, the incorporation of 7.5 wt. % RHA leads to a reduction of about 24 °C and 14°C of the temperatures where 5% and 50% weight loss was recorded, respectively, suggesting that with the introduction of RHA, the degradation of the matrix started earlier when compared to the neat PBAT/PLA. Considering that RHA is a stable material since it contains about 90% of silica, an inorganic compound with high-temperature resistance, the reduction of temperatures of biocomposites, recorded in 5% and 50% weight loss should be attributed to the hydrophilic character of RHA [19, 20].

Table 3. Characteristic values determined from TGA for neat PBAT/PLA and its biocomposites.

Materials	T_{onset} (°C)	$T_{5\%}$ [a] (°C)	$T_{50\%}$ [b] (°C)	T_{max} (°C)	RWP[c] at 180°C (%)	RWF[d] at 600°C (%)
Neat PBAT/PLA	304	334.7	370	371.2	99.3	1.7
PBAT/PLA /RHA 2.5%	279.6	331.4	362.6	363.6	100	5.9

(Table 3) cont.....

Materials	T_{onset} (°C)	$T_{5\%}{}^{(a)}$ (°C)	$T_{50\%}{}^{(b)}$ (°C)	T_{max} (°C)	RWP[c] at 180°C (%)	RWF[d] at 600°C (%)
PBAT/PLA /RHA 5.0%	289	317.6	364	361.1	100	10.4
PBAT/PLA /RHA 7.5%	284.5	310.8	355.7	351.6	100	15.2

(a) temperature 5% weight loss, (b) temperature 50% weight loss, (c) residual weight (RWP) at highest temperature used during processing by FDM 3D printing (180 °C), (d) residual weight (RWF) at 600°C.

As can be noted in Table **3**, the residual weight of the neat PBAT/PLA and its biocomposites at 180 °C (the highest temperature used during processing by FDM 3D printing) was about 99% for neat PBAT/PLA and 100% for biocomposites, indicating that neat PBAT/PLA basically, was not decomposed at the temperature used in 3D print processing, while its biocomposites were not decomposed at 3D print processing temperatures equal to or smaller than180 °C. These results show that 180 °C is a suitable temperature for FDM 3D printing. In addition, the residual weight of the biocomposites at 600 °C increased with the addition of RHA content in the matrix, which indicates that the addition of RHA increased the thermal stability of the biocomposites, despite the T_{onset} of the biocomposites being lower when compared to neat PBAT/PLA. Thus, it is possible to infer that the thermal stability of the PBAT/PLA/RHA biocomposites makes them suitable for the development of material using processing by FDM 3D printing.

3.2. FTIR Analyses

FTIR-ATR spectra in the range of 4000 to 650 cm^{-1} for RHA, neat PBAT/PLA, and its biocomposites are shown in Fig. (**6**). The characteristic PBAT/PLA peaks appeared at 731, 1086, 1182, 1272, and 1456 cm^{-1} [21-24]. The peak at 731 cm^{-1} is attributed to methylene [-(CH2)4], and the elongation of C=H bonds present in the aromatic rings of PBAT. The peaks at 1086, 1182, and 1272 cm^{-1} correspond to symmetrical elongation of C=O bonds and asymmetrical elongation of C-O-C bonds present in PBAT and PLA [21, 22]. The peak at 1456 cm^{-1} was attributed to the asymmetric and symmetrical elongation of CH_3 bonds in PBAT and PLA [22-24]. Based on FTIR spectra, it can be inferred that the rice husk ash corresponds to the majority of the silica-based minerals [25]. The RHA spectra present two main peaks, a strong one at 1096 cm^{-1} and one at 795 cm^{-1} related to asymmetric and symmetric stretching mode of SiO-Si [25, 26].

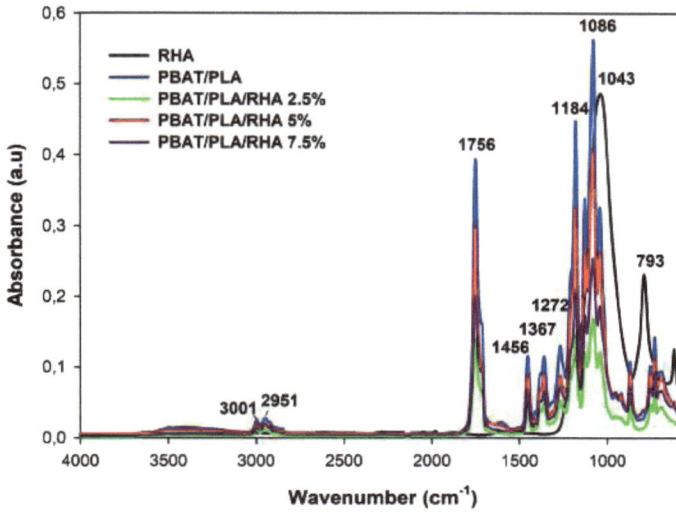

Fig. (6). FTIR-ATR spectra of RHA, neat PBAT/PLA, and its biocomposites.

3.3. SEM Analyses

SEM images of neat PBAT/PLA and its biocomposites are presented in Fig. (**7**). Fig. (**7a**) shows the image of neat PBAT/PLA with magnifications of 1.500 x, which present a characteristic morphology of immiscible blends showing the PLA phase dispersion in the PBAT matrix. The SEM image of PBAT/PLA/RHA with 2.5 wt. % RHA (Fig. **7b**) exhibits a rough surface with micrometric particles of the RHA of different sizes homogeneously distributed within the polymeric matrix. In addition, this figure also reveals the absence of aggregates of the filler on the matrix, suggesting good adhesion between RHA and PBAT/PLA. Unlike the biocomposite with 2.5% by weight of RHA, in the biocomposite containing 5 wt. % of RHA (Fig. **7c**), filler aggregates, holes and cavities due to extraction of RHA particles from the matrix are clearly visible. In addition, phase separations at the interface between the RHA domains and the matrix are also observed for this biocomposite. These suggest a poor bonding between the RHA and matrix, and also poor dispersion of RHA particles when the RHA content is about 5 wt. %. For biocomposites containing 7.5 wt.% of RHA (Fig. **7d**), a similar trend was observed, which showed numerous large holes and cavities and a higher extraction of RHA than the PBAT/PLA/RHA with 5 wt.% of RHA (Fig. **7c**).

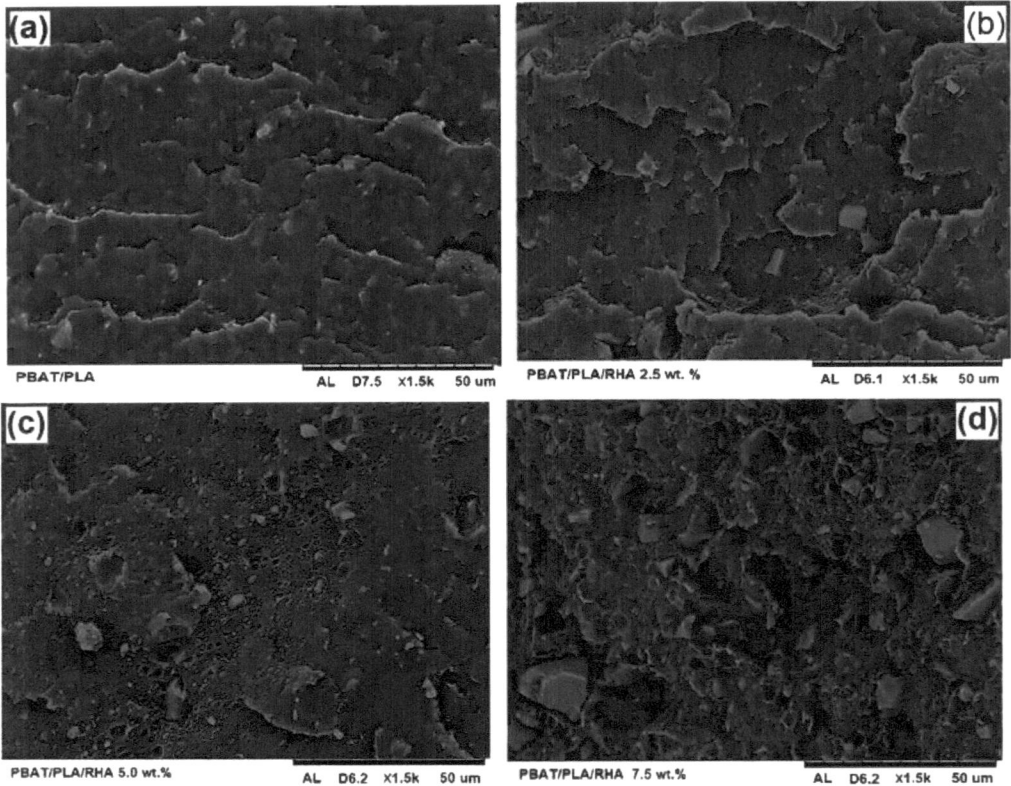

Fig. (7). SEM images of (**a**) neat PBAT/PLA, (**b**) PBAT/PLA/RHA 2.5 wt.% (**c**) PBAT/PLA/RHA 5.0 wt.%
(**d**) PBAT/PLA/RHA 7.5 wt.%.

Considering that for better interactions between the filler and matrix, the
wettability of the filler particles present in the matrix must be efficient. The
inadequate RHA (hydrophilic) wetting because of the difference in polarity
related to the PBAT/PLA (hydrophobic) might have resulted in the inefficient
wettability of the RHA surface. The poor wetting led to the particles
agglomeration instead of being uniformly distributed in the melt [27]. This
obviously caused a poor interface between the RHA and PBAT/PLA with the
presence of the agglomerates and RHA being extracted was observed in the
micrographs. Tipachan *et al.* [28] recently reported similar results. They observed
numerous holes and cavities on the surface of the PLA/RHA with increased
loading levels (5, 10 wt.%) of RHA in PLA, caused by RHA extraction from the
matrix due to poor interfacial adhesion between PLA and RHA.

3.4. Mechanical Tests

Table **4** summarizes the tensile test results of the neat PBAT/PLA and its biocomposites printed by FDM with 100% infill pattern and with varying orientation angles of 0, 45, and 90-degree. It can be inferred from Table **4** that the tensile properties of built parts from PBAT/PLA and its biocomposites were significantly influenced by the build orientation angle of parts.

As shown in Table **4**, the specimens printed with a 0° orientation presented higher tensile strength than the ones with a 45° and 90°. This is because the tensile strength of the specimens with a 0° orientation mainly depends on the strength of the filler, while the ones printed with 45° and 90° depend on the adhesion between the layers. Similar mechanical behaviors have been previously reported [29, 30].

Table 4. Mechanical tests results of the neat PBAT/PLA and its biocomposites.

Materials	$\sigma^{(a)}$ 0° (MPa)	$\sigma^{(a)}$ 45° (MPa)	$\sigma^{(a)}$ 90° (MPa)	$\varepsilon^{(b)}$ 0° (%)	$\varepsilon^{(b)}$ 45° (%)	$\varepsilon^{(b)}$ 90° (%)
Neat PBAT/PLA	33.6±1.3	33.1±1.3	32.2±1.6	16.9±1.2	16.3±0.7	15.4±1.4
PBAT/PLA /RHA 2.5%	38.0±1.4	33.4±1.7	26.5±0.4	14.4±0.9	16.2±0.6	15.9±1.6
PBAT/PLA /RHA 5.0%	22.7±2.7	23.8±2.1	23.8±0.4	10.7±0.4	12.6±1.2	12.0±0.8
PBAT/PLA /RHA 7.5%	21.4±1.0	23.0±0.7	20.2±0.7	10.7±0.5	12.9±2.2	12.5±0.6

(a) tensile stress at break; (b) elongation at break

The RHA, when infused into PBAT/PLA at low proportions, like 2.5 wt.%, enhanced the stiffness and the toughness of the printed specimens. The printed specimens of biocomposites containing 2.5wt.% of RHA, at 0° orientation, presented a slight enhancement of about 13% in tensile strength and a decrease of about 15% in elongation at break, compared to neat PBAT/PLA. This finding confirmed the good interaction between PBAT/PLA and RHA for the biocomposites with 2.5wt.% of filler, due to the homogeneous distribution of micrometric particles of the RHA within the matrix, which was observed by SEM image of PBAT/PLA/RHA containing 2.5wt.% of RHA (Fig. **7b**). Considering the toughness, with the increase of RHA contents into PBAT/PLA the material rigidity also increased. On the other hand, with the increase in the amount of RHA in the biocomposites to 5.0 wt. % and 7.5 wt. %, numerous holes, and cavities were observed due to the extraction of RHA particles from the matrix. Furthermore, agglomerates also appeared which eventually caused the deterioration of mechanical properties. For example, with 7.5 wt% of RHA in PBAT/PLA, the tensile strength, and elongation at break recorded was about 60% of the value of the neat PBAT/PLA. Similar outcomes were previously observed [29, 31].

CONCLUSION

In the present work, PBAT/PLA/carbonaceous biocomposites and filaments of neat PBAT/PLA and biocomposites, for 3D printing by FDM, were produced by the melt extrusion process from the incorporation of different loadings of the rice husk ash (RHA) into neat PBAT/PLA. Experimental investigations of tensile characteristics of specimens of neat PBAT/PLA and its biocomposites, printed by FDM 3D have been conducted. In addition, FTIR, TGA, and SEM analyses of neat PBAT/PLA and its biocomposites were performed. TGA analysis data revealed that the RHA contents increased the thermal stability of the biocomposites, as the residual weight of PBAT/PLA at 600°C increased with the incorporation of RHA. The reduction observed for the biocomposites at temperatures where 5% and 50% weight loss are recorded, was due to the hydrophilic character of RHA. The suitable temperature for FDM 3D printing was 180 °C according to TGA analyses data, as the residual weight of neat PBAT/PLA and its biocomposites at 180°C was about 99% and 100%, respectively. SEM image of the PBAT/PLA/RHA with 2.5 wt. % RHA exhibits a rough surface structure with a homogeneous distribution of RHA into the matrix, and without aggregates, suggesting that for 2.5wt.% RHA contents the adhesion between RHA and PBAT/PLA was good. In contrast, the biocomposite containing 5 wt.% and 7.5 wt.% of RHA reveals filler aggregates, holes, and cavities due to RHA particles extricated from the PBAT/PLA, suggesting a poor adhesion between the two phases, and also poor dispersion of RHA into the matrix. Considering the toughness, the introduction of RHA into PBAT/PLA increased the material rigidity. The tensile strength of specimens printed with 0° orientation was higher than the ones with 45° and 90°. This indicated that the tensile properties of materials were significantly influenced by the build orientation angle of specimens. According to the results, the application of the FDM 3D printing process, a rapid and low-cost fabrication, in carbonaceous biocomposites is suitable to predict the tensile properties of biocomposites in development, instead of using time-consuming and expensive manufacturing and testing. FDM 3D printing involves a large number of parameters, which may have effects on the properties of the resultant printed biocomposites. Thus, future investigations are still required for a more comprehensive evaluation of the correlations between structure and the properties of printed carbonaceous biocomposites.

FUNDING SOURCES

The authors would like to thank the funding from FAPESP/ Process Number 2019/00862-9.

ACKNOWLEDGEMENT

The authors wish to thank FAPESP, CAPES and CNPq to provide support for this work.

REFERENCES

[1]   Dios, J.R.; García-Astrain, C.; Costa, P.; Viana, J.C.; Lanceros-Méndez, S. Carbonaceous Filler Type and Content Dependence of the Physical-Chemical and Electromechanical Properties of Thermoplastic Elastomer Polymer Composites. *Materials (Basel),* **2019,** *12*(9), 1405.
        [http://dx.doi.org/10.3390/ma12091405] [PMID: 31052175]

[2]   Kim, K.W.; Han, W.; Kim, B.J. Effects of Mixing Ratio of Hybrid Carbonaceous Fillers on Thermal Conductivity and Mechanical Properties of Polypropylene Matrix Composites. *Polymers (Basel),* **2022,** *14*(10), 1935.
        [http://dx.doi.org/10.3390/polym14101935] [PMID: 35631820]

[3]   Benny, L.; John, A.; Varghese, A.; Hegde, G.; George, L. Waste elimination to porous carbonaceous materials for the application of electrochemical sensors: Recent developments. *J. Clean. Prod.,* **2021,** *290*, 125759.
        [http://dx.doi.org/10.1016/j.jclepro.2020.125759]

[4]   Yang, D.P.; Li, Z.; Liu, M.; Zhang, X.; Chen, Y.; Xue, H.; Ye, E.; Luque, R. Biomass-Derived Carbonaceous Materials: Recent Progress in Synthetic Approaches, Advantages, and Applications. *ACS Sustain. Chem.& Eng.,* **2019,** *7*(5), 4564-4585.
        [http://dx.doi.org/10.1021/acssuschemeng.8b06030]

[5]   Dey, A.; Yodo, N. A systematic survey of fdm process parameter optimization and their influence on part characteristics. *Journal of Manufacturing and Materials Processing,* **2019,** *3*(3), 64.
        [http://dx.doi.org/10.3390/jmmp3030064]

[6]   Wickramasinghe, S.; Do, T.; Tran, P. Fdm-based 3d printing of polymer and associated composite: a review on mechanical properties, defects and treatments. *Polymers (Basel),* **2020,** *12*(7), 1529.
        [http://dx.doi.org/10.3390/polym12071529] [PMID: 32664374]

[7]   Luo, J.; Wang, H.; Zuo, D.; Ji, A.; Liu, Y. Research on the Application of MWCNTs/PLA Composite Material in the Manufacturing of Conductive Composite Products in 3D Printing. *Micromachines (Basel),* **2018,** *9*(12), 635.
        [http://dx.doi.org/10.3390/mi9120635] [PMID: 30513580]

[8]   Liu, Z.; Wang, Y.; Wu, B.; Cui, C.; Guo, Y.; Yan, C. A critical review of fused deposition modeling 3D printing technology in manufacturing polylactic acid parts. *Int. J. Adv. Manuf. Technol.,* **2019,** *102*(9-12), 2877-2889.
        [http://dx.doi.org/10.1007/s00170-019-03332-x]

[9]   Jagadeesh, P.; Puttegowda, M.; Rangappa, S.M.; Alexey, K.; Gorbatyuk, S.; Khan, A.; Doddamani, M.; Siengchin, S. A comprehensive review on 3D printing advancements in polymer composites: technologies, materials, and applications. *Int. J. Adv. Manuf. Technol.,* **2022,** *121*(1-2), 127-169.
        [http://dx.doi.org/10.1007/s00170-022-09406-7]

[10]   Sciancalepore, C.; Togliatti, E.; Marozzi, M.; Rizzi, F.M.A.; Pugliese, D.; Cavazza, A.; Pitirollo, O.; Grimaldi, M.; Milanese, D. Flexible PBAT-Based Composite Filaments for Tunable FDM 3D Printing. *ACS Appl. Bio Mater.,* **2022,** *5*(7), 3219-3229.
        [http://dx.doi.org/10.1021/acsabm.2c00203] [PMID: 35729847]

[11]   Mani, M.; Karthikeyan, A.G.; Kalaiselvan, K.; Muthusamy, P.; Muruganandhan, P. Optimization of FDM 3-D printer process parameters for surface roughness and mechanical properties using PLA material. *Mater. Today Proc.,* **2022,** *66*, 1926-1931.
        [http://dx.doi.org/10.1016/j.matpr.2022.05.422]

[12] Liou, T.H.; Wang, P.Y. Utilization of rice husk wastes in synthesis of graphene oxide-based carbonaceous nanocomposites. *Waste Manag.,* **2020**, *108,* 51-61.
[http://dx.doi.org/10.1016/j.wasman.2020.04.029] [PMID: 32344300]

[13] Herrera, K.; Morales, L.F.; Tarazona, N.A.; Aguado, R.; Saldarriaga, J.F. Use of Biochar from Rice Husk Pyrolysis: Part A: Recovery as an Adsorbent in the Removal of Emerging Compounds. *ACS Omega,* **2022**, *7*(9), 7625-7637.
[http://dx.doi.org/10.1021/acsomega.1c06147] [PMID: 35284759]

[14] Sharma, S.K.; Sharma, G.; Sharma, A.; Bhardwaj, K.; Preeti, K.; Singh, K.; Kumar, A.; Pal, V.K.; Choi, E.H.; Singh, S.P.; Kaushik, N.K. Synthesis of silica and carbon-based nanomaterials from rice husk ash by ambient fiery and furnace sweltering using a chemical method. *Applied Surface Science Advances,* **2022**, *8,* 100225.
[http://dx.doi.org/10.1016/j.apsadv.2022.100225]

[15] Bushra, B.; Remya, N. Biochar from pyrolysis of rice husk biomass—characteristics, modification and environmental application. *Biomass Convers. Biorefin.,* **2020**.
[http://dx.doi.org/10.1007/s13399-020-01092-3]

[16] Gan, P.G.; Sam, S.T.; Abdullah, M.F.; Omar, M.F. Thermal properties of nanocellulose☐reinforced composites: A review. *J. Appl. Polym. Sci.,* **2020**, *137*(11), 48544.
[http://dx.doi.org/10.1002/app.48544]

[17] Andrade, M.S.; Ishikawa, O.H.; Costa, R.S.; Seixas, M.V.S.; Rodrigues, R.C.L.B.; Moura, E.A.B. Development of sustainable food packaging material based on biodegradable polymer reinforced with cellulose nanocrystals. *Food Packag. Shelf Life,* **2022**, *31,* 100807.
[http://dx.doi.org/10.1016/j.fpsl.2021.100807]

[18] Shrivastava, N.K.; Wooi, O.S.; Hassan, A.; Inuwa, I.M. Mechanical and flammability properties of poly(lactic acid)/poly(butylene adipate-co-terephthalate) blends and nanocomposites: Effects of compatibilizer and graphene. *Malaysian Journal of Fundamental and Applied Sciences,* **2018**, *14*(4), 425-431.
[http://dx.doi.org/10.11113/mjfas.v14n4.1233]

[19] Zuo, Y.; Chen, K.; Li, P.; He, X.; Li, W.; Wu, Y. Effect of nano-SiO_2 on the compatibility interface and properties of polylactic acid-grafted-bamboo fiber/polylactic acid composite. *Int. J. Biol. Macromol.,* **2020**, *157,* 177-186.
[http://dx.doi.org/10.1016/j.ijbiomac.2020.04.205] [PMID: 32344087]

[20] Venkatesan, R.; Rajeswari, N. Preparation, Mechanical and Antimicrobial Properties of SiO2/ Poly(butylene adipate-co-terephthalate) Films for Active Food Packaging. *Silicon,* **2019**, *11*(5), 2233-2239.
[http://dx.doi.org/10.1007/s12633-015-9402-8]

[21] Facchi, D.P.; Souza, P.R.; Almeida, V.C.; Bonafé, E.G.; Martins, A.F. Optimizing the Ecovio® and Ecovio®/zein solution parameters to achieve electrospinnability and provide thin fibers. *J. Mol. Liq.,* **2021**, *321,* 114476.
[http://dx.doi.org/10.1016/j.molliq.2020.114476]

[22] Barros, J.J.P.; Soares, C.P.; Moura, E.A.B.; Wellen, R.M.R. Enhanced miscibility of PBAT/PLA/lignin upon γ ☐irradiation and effects on the non☐isothermal crystallization. *J. Appl. Polym. Sci.,* **2022**, *139*(45), e53124.
[http://dx.doi.org/10.1002/app.53124]

[23] Carvalho, B.M.; Pellá, M.C.G.; Hardt, J.C.; de Souza Rossin, A.R.; Tonet, A.; Ilipronti, T.; Caetano, J.; Dragunski, D.C. Ecovio®-based nanofibers as a potential fast transdermal releaser of aceclofenac. *J. Mol. Liq.,* **2021**, *325,* 115206.
[http://dx.doi.org/10.1016/j.molliq.2020.115206]

[24] Wang, L.F.; Rhim, J.W.; Hong, S.I. Preparation of poly(lactide)/poly(butylene adipate-c--terephthalate) blend films using a solvent casting method and their food packaging application.

Lebensm. Wiss. Technol., **2016**, *68*, 454-461.
[http://dx.doi.org/10.1016/j.lwt.2015.12.062]

[25] Běhálek, L.; Borůvka, M.; Brdlík, P.; Habr, J.; Lenfeld, P.; Kroisová, D.; Veselka, F.; Novák, J. Thermal properties and non-isothermal crystallization kinetics of biocomposites based on poly(lactic acid), rice husks and cellulose fibres. *J. Therm. Anal. Calorim.,* **2020**, *142*(2), 629-649.
[http://dx.doi.org/10.1007/s10973-020-09894-3]

[26] Amutha, K.; Ravibaskar, R.; Sivakumar, G. Extraction, Synthesis and Characterization of Nanosilica from Rice Husk Ash. *Int. J. Nanotechnol. Appl.,* **2010**, *4*, 61.

[27] Ayswarya, E.P.; Vidya Francis, K.F.; Renju, V.S.; Thachil, E.T. Rice husk ash – A valuable reinforcement for high density polyethylene. *Mater. Des.,* **2012**, *41*, 1-7.
[http://dx.doi.org/10.1016/j.matdes.2012.04.035]

[28] Tipachan, C.; Gupta, R.K.; Agarwal, S.; Kajorncheappunngam, S. Flame retardant properties and thermal stability of polylactic acid filled with layered double hydroxide and rice husk ash silica. *J. Polym. Environ.,* **2020**, *28*(3), 948-961.
[http://dx.doi.org/10.1007/s10924-020-01658-2]

[29] Ding, Q.; Li, X.; Zhang, D.; Zhao, G.; Sun, Z. Anisotropy of poly(lactic acid)/carbon fiber composites prepared by fused deposition modeling. *J. Appl. Polym. Sci.,* **2020**, *137*(23), 48786.
[http://dx.doi.org/10.1002/app.48786]

[30] Ziemian, S.; Okwara, M.; Ziemian, C.W. Tensile and fatigue behavior of layered acrylonitrile butadiene styrene. *Rapid Prototyping J.,* **2015**, *21*(3), 270-278.
[http://dx.doi.org/10.1108/RPJ-09-2013-0086]

[31] Ning, F.; Cong, W.; Qiu, J.; Wei, J.; Wang, S. Additive manufacturing of carbon fiber reinforced thermoplastic composites using fused deposition modeling. *Compos., Part B Eng.,* **2015**, *80*, 369-378.
[http://dx.doi.org/10.1016/j.compositesb.2015.06.013]

CHAPTER 7

Biochar-Based Polymer Composites: A Pathway to Enhanced Electrical Conductivity

Mahesh K. Pallikonda[1,*] and **Joao A. Antonangelo**[2]

[1] *Department of Engineering Technology, Austin Peay State University, Clarksville, TN 37044, USA*

[2] *Department of Crop and Soil Sciences, Washington State University , Pullman, WA 99164, USA*

Abstract: In the past 20-25 years, biochar has been promoted as a valuable resource of a carbon filler in polymer composites, sustainable agriculture, and environmental quality protection given its improved porous structure and electrochemical properties in comparison to other carbon-based materials. Recent works focusing on biochar and biochar-based nanocomposites are highlighting such properties and are even enhanced with nanotechnology. The higher porosity attributed to biochar is highlighted along with its great electrochemical properties able to retain nutrients for longer and favors their slow release. The use of biochar as a filler material to improve the electrical conductivity properties of polymers and the emphasis on various parameters, such as pyrolysis temperature, the type of feedstock, and compaction pressures on the electrical conductivity of the resultant composites are discussed.

Keywords: Biochar, Electrical Conductivity, Polymer Composites, Percolation, Organic Conductors.

1. INTRODUCTION

Biochar (BC) is a charcoal-based material made from biomass, such as wood, crop residues, or other organic materials. It is produced by pyrolyzing (heating in the absence of oxygen) those materials [1, 2]. The pyrolysis process can be carried out in a variety of methods, including using a kiln, retort, or gasifier [3 - 5]. The pyrolysis temperature and duration, as well as the type of biomass used, can affect the properties of the resulting BC. Additionally, the production of BC has positive effects on the environment. It sequesters carbon from the atmosphere and stores it in the soil, reducing greenhouse gas emissions and improving soil health [6, 7].

* **Corresponding author Mahesh K. Pallikonda:** Department of Engineering Technology, Austin Peay State University, Clarksville, TN 37044, USA; Tel: +1 931 221 1484; E-mail: pallikondam@apsu.edu

Deepa Kodali & Vijaya Rangari (Eds.)

Furthermore, it can also help with waste management [3, 8], as it can be produced from agricultural waste and other organic materials that would otherwise be discarded. Historically, BC has been recognized as a valuable product in agriculture since it contributes to crop yield enhancement, serves as an organic fertilizer, neutralizes soil pH, and supports media for microbial growth [9, 10]. In the soil, biochar was also found to reduce lime application because of its potential to neutralize the pH of acidic soils [11 - 13] and minimize the application of commercial fertilizers [14]. In addition to increasing the available levels of macronutrients such as nitrogen (N), phosphorus (P) and potassium (K), biochar addition also reduces the need for nitrogen (N) application, which in turn reduces N leaching to groundwater [15]. The role of biochar in increasing soil pH and soil organic carbon (SOC) storage, as well as improving cation exchange capacity (CEC) was also highlighted in the past literature [16].

Due to sustainable production methods and ease of production, researchers consider BC an alternative material filler material in polymers and ceramics [17 - 19]. Carbon-based filler materials such as Carbon nanotubes (CNTs), Graphene, and Carbon black are already established as filler materials in various materials to produce selectively engineered products. Products engineered for specific applications include conductive polymers, high conductive alloys, and reinforced alloys for special applications such as high corrosion resistance, high strength and temperature resistance in polymers, and others [20 - 23]. The carbon structure of biochar can be characterized by several parameters such as pore size, the intensity of pores, ash content, pH value, and others.

Among many special applications, materials with low electrical resistance have special attention among researchers in the fields of electronics and material science. The growing need for special sensors and flexible electronics created an opportunity to exploit suitable materials for organic conductors. Organic conductors are conductive polymers developed by embedding carbon fillers [24 - 27]. Carbon fillers such as CNTs and Graphene showed promising results, however, the cost of these fillers is relatively high compared to BC. The carbon structure of BC makes it an attractive candidate for use in polymer composites owing to its high electrical conductivity and thermal stability. BC can be added to polymers such as polyaniline (PANI), Polypropylene (PP), and Polyvinyl Alcohol (PVA) to enhance their electrical conductivity, which promotes the potential to be used in areas such as energy storage and electronic devices. However, additional research is necessary to fully clarify the potential of BC in this area and to develop practical and cost-effective methods for incorporating it into polymers. In addition, not all BCs are the same, different types of BCs will have different properties, and the specific application will determine which type of BC is the most appropriate. There are a multitude of factors that affect the properties of resulting BC, the type of production method, the type and condition of feedstock (biomass), the temperature of pyrolysis, and others [28, 29].

The electrical conductivity presented in biochar-filled polymers is significantly affected by the temperature of pyrolysis used during the biochar production. High-pyrolysis temperatures usually yield biochar with higher carbon content, which can result in higher electrical conductivity when used as a filler in polymers. This is because the carbon structure of biochar is highly porous, and the pores can act as pathways for electrical current to flow. Higher carbon content in the biochar will result in more pores and therefore more pathways for electrical current to flow, increasing conductivity. Additionally, at higher temperatures, the biochar structure will be more graphitized, meaning that the carbon atoms are more ordered and the electrical conductivity will be higher. On the other hand, low-pyrolysis temperatures tend to produce biochar with lower carbon content, which can result in lower electrical conductivity when used as a filler in polymers. It is worth noting that the electrical conductivity of biochar-filled polymers is also affected by the polymer type and the biochar loading in the polymer matrix. This paper discusses the mechanisms of electrical conductivity in polymers followed by experimental studies on the factors that affect electrical conductivity in biochar-filled polymer composites.

2. MECHANISMS OF ELECTRICAL CONDUCTIVITY IN POLYMER COMPOSITES

The theory of electrical conductivity in polymers is based on the idea that polymers are made up of long chains of repeating units called monomers, which are held together by covalent bonds. Monomer chains have various levels of structures and can be made up of complete amorphous to varied ranges of crystalline.

In general, electrical conductivity in polymers arises from the movement of charged particles such as electrons, holes, or ions within their matrix. The movement of these particles can be caused by several mechanisms [30 - 33], such as hopping, tunneling, and percolation as shown in Fig. (1).

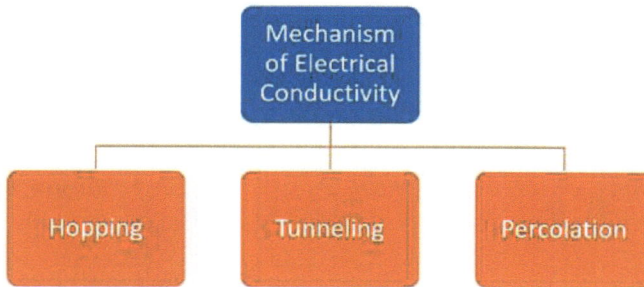

Fig. (1). Various types of electron transport mechanisms.

Hopping is a common mechanism for electrical conductivity in amorphous polymers, where charge carriers move between localized sites due to the differences in thermal energy [31, 33]. Tunneling is another mechanism in ordered or crystalline polymers, where the charge carriers can tunnel through the regions of the polymer matrix [30, 31]. Percolation is observed in polymers that contain conductive filler particles, where the particles form a network that spans the polymer matrix and allows for the movement of charge carriers [32, 34].

In carbon-based material-filled polymers, the electrical conductivity is achieved through the percolation mechanism [32]. Percolation is the process of forming conductive pathways between the carbon particles in the polymer matrix. The formation of these pathways occurs when the carbon particles reach a critical concentration, known as the percolation threshold [34].

Below the percolation threshold, the carbon particles are not in contact with each other and the polymer matrix behaves as an insulator. However, when the concentration of carbon particles exceeds the percolation threshold, the particles begin to touch each other, forming a continuous network of conductive pathways throughout the polymer matrix [34]. This leads to an increase in electrical conductivity, as the electrons can now move freely through the conductive network.

The percolation threshold and the resulting electrical conductivity in carbon-filled polymers depend on various factors, including the shape and size of the carbon particles, their distribution within the polymer matrix, and the properties of the polymer itself [31, 34].

The percolation threshold can be quantified by measuring the electrical conductivity of a polymer composite as a function of the volume fraction of conductive filler [34]. At low filler concentrations, the conductivity is typically low and shows little dependence on the filler content. As the filler concentration is increased, there is a sudden increase in the conductivity at a certain critical filler concentration, which is known as the percolation threshold. The percolation threshold can be estimated by fitting a power law to the conductivity data and extrapolating to the point where the exponent reaches a critical value, as shown in Fig. (**2**). Alternatively, it can be determined by measuring the electrical conductivity of the composite at various filler concentrations and finding the concentration at which the conductivity starts to increase rapidly [30, 32 - 35].

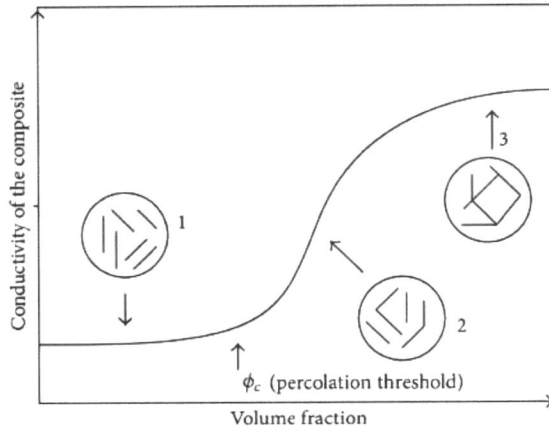

Fig. (2). Trends of Electrical conductivity before, at, and after percolation threshold of volume fraction of filler concentration. [copyright © 2013 Rafael Vargas-Bernal *et al.* [35]].

$$\sigma = \sigma_f\left(\varphi_f - \varphi_p\right)^b$$

σ_f = Conductivity of the filler material

φ_f = Volume fraction of the filler in the polymer matrix

φ_p = Percolation threshold of the polymer

b = Exponent of the nanocomposite material with reported values ranging from 1.3 to 3.1 by existing studies.

However, percolation theory alone cannot justify the electron transport in the biochar-filled polymers. Filler material often comes in various sizes and carbon content [30 - 35]. There is a plethora of factors that influence the overall conductivity of the polymer. Overall, there are studies that reported biochar-filled polymers that exhibit properties that align with the percolation theory and there are also studies reported that do not follow the percolation theory. Moreover, it is noticed that the percolation behavior of biochar-filled polymers can vary depending on factors such as the type of polymer matrix, the type and characteristics of the biochar filler, and the processing conditions used to prepare the composites.

3. FACTORS INFLUENCING THE ELECTRICAL CONDUCTIVITY IN POLYMER COMPOSITES

The mechanisms of electron transport in polymer composites are still actively investigated by researchers from various fields [30 - 35]. There are a multitude of factors that impact the flow of electrons. Polymers reinforced with biochar is a relatively new concept and no significant literature is available in the area focusing on the electrical conductivity of such materials. Based on the existing literature, there are multiple factors that affect electrical conductivity. This chapter focused on a few studies that fall under categories such as pyrolysis temperature to produce Biochar, degree of graphitization, raw material used as a precursor material to produce biochar, and concentration of biochar in the polymer matrix [36 - 43], as shown in Fig. (**3**).

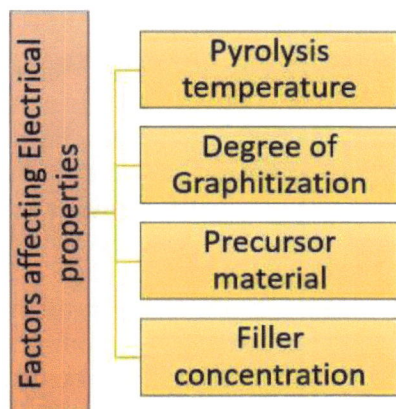

Fig. (3). Predominant factors affecting the electrical conductivity presented in this chapter.

3.1. Pyrolysis Temperature

Li *et al.* [36], conducted experimental studies to show the ability to use BC as a filler in ultra-high molecular weight polyethylene (UHMWPE). They showed encouraging results by using three BC materials produced from pine, apple, and Bamboo charcoal. In addition, EC measurements of BC material as well as BC-filled UHMWPE are measured for BC produced at 500, 700, 900, and 1100 C pyrolysis temperatures. Researchers reported the higher the carbonization temperature, the better the EC values. The polymer composites are produced by mixing 70% (w/w) BC with UHMWPE. Furthermore, to prove the ability of polymer composites' conductivity, researchers demonstrated a circuit using the extruded polymer composites to light a red light-emitting diode [36].

Giorcelli *et al*. [37], studied the properties of BC produced from Miscanthus at three different temperatures, *i.e.*, 650°C, 700°C, and 750°C. In this experimental study, researchers analyzed the samples produced at three pyrolysis temperatures which are followed by a CO_2 activation process. Electrical conductivity measurements are recorded on the compressed BC powder at different compression loadings. Reported pressures range from 200 to 1600 bar with 200 bar increments. The three samples MIS 650, MIS 700, and MIS 750 [37] produced at high–pyrolysis temperatures such as 650°C, 700°C, and 750°C respectively. Overall, from the pressure ranges it is evident that MIS 750 showed higher EC when compared with the other two BC materials. The authors justified the results with a Raman analysis that showed higher carbon content in MIS 750 when compared with the other two parts.

3.2. Degree of Graphitization

In later studies, researchers also confirmed the formation of graphitic structures in BC at heat treatments above 900°C, which showed improved electrical conductivity. Gabhi *et al*. [38], conducted experimental studies to understand the EC of BC by observing the relationship between BC's structure and property. In this study, the researchers reported that the bulk conductivity of BC is directly proportional to the carbon content. They observed an increase in EC from 2.5×10^{-4} to 399.7 S/m when carbon content increased from 86.8 to 93.7 wt%. They also observed upon heat treatment, the EC improved dramatically. Researchers reported that heat treatment at 950°C for 8 hours improved EC from 0.16 to 106.41 S/m. They observed upon heat treatment that there was the formation of graphitic structures and the degree of graphitization together with the degree of carbon content improved the EC of heat-treated BC.

3.3. Precursor Material

Kane *et al*. [39] studied the role of raw materials used for producing Biochar in electrical conductivity. For this study, they used raw materials predominantly based on Lignin, cellulose, and biomass. Furthermore, they studied three different forms of lignin-based raw materials. In their experimental studies, it is observed that biochar produced from lignin-based raw material showed the highest electrical conductivity at a pyrolysis temperature of 1100°C. There was a structural change in the biomass between the material synthesized at 900°C and 1100°C pyrolysis temperatures.

Rehrah *et al.* [40] investigated various raw materials as the precursors for biochar and noticed similar results. They noticed EC values ranging from 815μScm^{-1} to 1971 μScm^{-1} in biochar produced from cotton gin at increasing pyrolysis temperature and time. On the other hand, Biochar produced from pecan shells exhibited lower EC. Researchers attributed their findings to the increasing ash content as the main ingredient in higher EC. It is noticed in their experimental studies that cotton gin has higher ash content compared with pecan shells. Additionally, as the pyrolysis temperature increased, higher ash content was noticed. This is in correlation to their findings of EC, which increased with increased pyrolysis temperature for various precursor materials.

3.4. Different Concentrations of Filler Material

Nan *et al.* [41], studied the behavior of incorporating different w% of mixed hardwood BC into PVA. They investigated the EC of incorporating 2%, 6%, and 10% (w/w) BC into a 10% solution of PVA which was dried and EC measurements were reported. They noticed no significant EC at 2 wt%, however, at 6 wt% and 10 wt% a significant improvement was noticed. In addition, EC of 0.238 x 10-6 S/cm and 1.833 x 10-6 S/cm for 6 wt% and 10 wt%, was noticed respectively. They compared their findings to the established results of graphene-filled PVA at similar wt%. Researchers concluded the findings for both BC-filled PVA and Graphene-filled PVA showed a similar range of EC.

Poulose *et al.* [42], used date palm waste to produce BC at two high–pyrolysis temperatures, 700°C and 900°C. This BC is mixed with PP to study the behavior of the resultant polymer composite. It is observed that the EC of the BC/PP composite improved fourfold upon increasing the BC content from 0 to 15% wt%. Although an improvement is noticed, the authors also highlighted that the improved conductivity is not as significant as it was observed in carbon blank-filled PP composites. Researchers attribute this to the formation of agglomerations and the inability to form a continuous BC/PP system as one of the reasons for the lower conductivity [42]. Although authors reported agglomerations as one of the reasons for low conductivity, it is important to note that the formation of agglomerations is an infamous tendency of carbon-based filler materials. Additionally, the second is credited to the high ash content of BC, which inhibits the formation of continuous conductive networks.

Li *et al* [43] synthesized an 80 wt% biochar-filled UHMWPE/LLDPE nanocomposite using extrusion and hot compression methods. In their investigation, they showed a remarkable increase in the electrical conductivity of the material. Biochar used for this study was prepared by carbonizing bamboo at 1100°C. It is reported that an electrical conductivity of 107.6 S/m is exhibited by this nanocomposite. Additionally, they found at 60wt% the electrical conductivity was 5.3 S/m. Researchers attributed the electron transport in the material to the physical contact of the biochar in the polymer matrix and a special tunneling mechanism in the instances where the biochar particles are near each other, thereby forming a conductive network.

CONCLUSION

This chapter has shown that the use of biochar in the synthesis of polymer nanocomposites as a filler improves the electrical conductivity. The electrical conductivity of biochar can depend not only on the type of pyrolysis temperature but also on the type of feedstock used to produce it and other factors such as surface area, pH, and elemental composition. Different feedstocks will have different carbon contents and structural properties, which can affect the conductivity of the resulting biochar. For example, biochar produced from wood feedstock may have a higher electrical conductivity than biochar produced from agricultural waste feedstock, due to the higher carbon content in wood. Additionally, the chemical composition of the feedstock will have an effect on the electrical conductivity of the biochar, for instance, biochar produced from lignocellulosic feedstocks (such as wood and agricultural waste) tends to have lower electrical conductivity than biochar produced from feedstocks rich in lignin (such as straw) due to the different chemical structure of these compounds. Furthermore, the characteristics of the feedstock, such as moisture content, ash content, or mineral content, can also affect the conductivity of the resulting biochar. It is also worth noting that the conductivity of biochar can also be affected by the pyrolysis temperature, duration, and the presence of catalysts. Therefore, in order to achieve the desired properties of biochar for a specific application, it is important to select the appropriate feedstock and optimize pyrolysis conditions to produce biochar with the desired conductivity.

REFERENCES

[1]   Nanda, S.; A. Dalai, K.; Berruti, F.; Kozinski A, J. "Biochar as an Exceptional Bioresource for Energy, Agronomy, Carbon Sequestration, Activated Carbon and Specialty Materials,". *Waste Biomass Valoriz.,* **2016,** *7,* 201-235. [Springer Netherlands.].

[2]   Das, O.; Bhattacharyya, D.; Sarmah, A.K. Sustainable eco–composites obtained from waste derived biochar: A consideration in performance properties, production costs, and environmental impact. *J. Clean. Prod.,* **2016,** *129,* 159-168.
[http://dx.doi.org/10.1016/j.jclepro.2016.04.088]

[3]   Panwar, N.L.; Pawar, A.; Salvi, B.L. Comprehensive review on production and utilization of biochar. *SN Applied Sciences,* **2019,** *1*(2), 168.
[http://dx.doi.org/10.1007/s42452-019-0172-6]

[4]   Park, J.Y.; Kim, J.K.; Oh, C.H.; Park, J.W.; Kwon, E.E. Production of bio-oil from fast pyrolysis of biomass using a pilot-scale circulating fluidized bed reactor and its characterization. *J. Environ.*

Manage., **2019**, *234*, 138-144.
[http://dx.doi.org/10.1016/j.jenvman.2018.12.104] [PMID: 30616185]

[5]   Umerah, C.O.; Kodali, D.; Head, S.; Jeelani, S.; Rangari, V.K. Synthesis of carbon from waste coconutshell and their application as filler in bioplast polymer filaments for 3D printing. *Compos., Part B Eng.,* **2020**, *202*, 108428.
[http://dx.doi.org/10.1016/j.compositesb.2020.108428]

[6]   Liu, H.; Kumar, V.; Yadav, V.; Guo, S.; Sarsaiya, S.; Binod, P.; Sindhu, R.; Xu, P.; Zhang, Z.; Pandey, A.; Kumar Awasthi, M. Bioengineered biochar as smart candidate for resource recovery toward circular bio-economy: A review. *Bioengineered,* **2021**, *12*(2), 10269-10301.
[http://dx.doi.org/10.1080/21655979.2021.1993536] [PMID: 34709979]

[7]   Singh Yadav, S.P.; Bhandari, S.; Bhatta, D.; Poudel, A.; Bhattarai, S.; Yadav, P.; Ghimire, N.; Paudel, P.; Paudel, P.; Shrestha, J.; Oli, B. Biochar application: A sustainable approach to improve soil health. *Journal of Agriculture and Food Research,* **2023**, *11*, 100498.
[http://dx.doi.org/10.1016/j.jafr.2023.100498]

[8]   Gupta, M.; Savla, N.; Pandit, C.; Pandit, S.; Gupta, P.K.; Pant, M.; Khilari, S.; Kumar, Y.; Agarwal, D.; Nair, R.R.; Thomas, D.; Thakur, V.K. Use of biomass-derived biochar in wastewater treatment and power production: A promising solution for a sustainable environment. *Sci. Total Environ.,* **2022**, *825*, 153892.
[http://dx.doi.org/10.1016/j.scitotenv.2022.153892] [PMID: 35181360]

[9]   Chausali, N.; Saxena, J.; Prasad, R. Nanobiochar and biochar based nanocomposites: Advances and applications. *Journal of Agriculture and Food Research,* **2021**, *5*, 100191.
[http://dx.doi.org/10.1016/j.jafr.2021.100191]

[10]   Wang, S.; Dai, G.; Yang, H.; Luo, Z. Lignocellulosic biomass pyrolysis mechanism: A state-of-the-art review. *Pror. Energy Combust. Sci.,* **2017**, *62*, 33-86.
[http://dx.doi.org/10.1016/j.pecs.2017.05.004]

[11]   Antonangelo, J.A.; Zhang, H. Heavy metal phytoavailability in a contaminated soil of northeastern Oklahoma as affected by biochar amendment. *Environ. Sci. Pollut. Res. Int.,* **2019**, *26*(32), 33582-33593.
[http://dx.doi.org/10.1007/s11356-019-06497-w] [PMID: 31586315]

[12]   Arthur Antonangelo, J.; Zhang, H. "The Use of Biochar as a Soil Amendment to Reduce Potentially Toxic Metals (PTMs) Phytoavailability," in Applications of Biochar for Environmental Safety. IntechOpen, **2020**.
[http://dx.doi.org/10.5772/intechopen.92611]

[13]   Antonangelo, J.A.; Zhang, H.; Sun, X.; Kumar, A. Physicochemical properties and morphology of biochars as affected by feedstock sources and pyrolysis temperatures. *Biochar,* **2019**, *1*(3), 325-336.
[http://dx.doi.org/10.1007/s42773-019-00028-z]

[14]   Acosta-Luque, M.P.; López, J.E.; Henao, N.; Zapata, D.; Giraldo, J.C.; Saldarriaga, J.F. Remediation of Pb-contaminated soil using biochar-based slow-release P fertilizer and biomonitoring employing bioindicators. *Sci. Rep.,* **2023**, *13*(1), 1657.
[http://dx.doi.org/10.1038/s41598-022-27043-8] [PMID: 36717659]

[15]   Li, D.; Manu, M.K.; Varjani, S.; Wong, J.W.C. Role of tobacco and bamboo biochar on food waste digestate co-composting: Nitrogen conservation, greenhouse gas emissions, and compost quality. *Waste Manag.,* **2023**, *156*, 44-54.
[http://dx.doi.org/10.1016/j.wasman.2022.10.022] [PMID: 36436407]

[16]   Chan, K.Y.; Van Zwieten, L.; Meszaros, I.; Downie, A.; Joseph, S. Using poultry litter biochars as soil amendments. *Soil Res.,* **2008**, *46*(5), 437.
[http://dx.doi.org/10.1071/SR08036]

[17]   Chakraborty, I.; Sathe, S.M.; Dubey, B.K.; Ghangrekar, M.M. Waste-derived biochar: Applications and future perspective in microbial fuel cells. *Bioresour. Technol.,* **2020**, *312*, 123587.

[http://dx.doi.org/10.1016/j.biortech.2020.123587] [PMID: 32480350]

[18] Zhang, Q.; Khan, M.U.; Lin, X.; Cai, H.; Lei, H. Temperature varied biochar as a reinforcing filler for high-density polyethylene composites. *Compos., Part B Eng.,* **2019**, *175*, 107151.
[http://dx.doi.org/10.1016/j.compositesb.2019.107151]

[19] Mandoreba, C.; Gwenzi, W.; Chaukura, N. Defluoridation of drinking water using a ceramic filter decorated with iron oxide-biochar composites. *Int. J. Appl. Ceram. Technol.,* **2021**, *18*(4), 1321-1329.
[http://dx.doi.org/10.1111/ijac.13768]

[20] Jeong, K.U.; Lim, J.Y.; Lee, J.Y.; Kang, S.L.; Nah, C. Polymer nanocomposites reinforced with multi-walled carbon nanotubes for semiconducting layers of high-voltage power cables. *Polym. Int.,* **2010**, *59*(1), 100-106.
[http://dx.doi.org/10.1002/pi.2696]

[21] Sreenivasulu, B.; Ramji, B.; Nagaral, M. A Review on Graphene Reinforced Polymer Matrix Composites. *Mater. Today Proc.,* **2018**, *5*(1), 2419-2428.
[http://dx.doi.org/10.1016/j.matpr.2017.11.021]

[22] Pallikonda, M.K.; Nayfeh, T.H. Experimental Investigation to Study the Feasibility of Fabricating Ultra-Conductive Copper Using a Hybrid Method. *Materials (Basel),* **2021**, *14*(19), 5560.
[http://dx.doi.org/10.3390/ma14195560] [PMID: 34639958]

[23] Pallikonda, M.K.P. Forming a metal matrix nanocomposite (mmnc) with fully dispersed and deagglomerated multiwalled carbon nanotubes (MWCNTs). In: *Master's thesis, Cleveland State University*;   ,   **2017**;   p.   79.   Available   from: http://rave.ohiolink.edu/ etdc/view?acc_num=csu1503937490966191

[24] Klauk, H. Organic Electronics: Materials, Manufacturing and Applications. Wiley-VCH Verlag GmbH & Co. KGaA **2006**, p. 446.
[http://dx.doi.org/10.1002/3527608753]

[25] Ishiguro, T.; Yamaji, K.; Saito, G. Organic Conductors. In: Organic Superconductors. Springer Series in Solid-State Sciences, Springer, Berlin, Heidelberg, *88*, **1998**, pp. 15-43.
[http://dx.doi.org/10.1007/978-3-642-58262-2_2]

[26] Shivakumar, S., Ed. The Flexible Electronics Opportunity. National Academies Press: Washington, D.C., **2014**.
[http://dx.doi.org/10.17226/18812]

[27] Sun, Y.; Rogers, J.A. Inorganic Semiconductors for Flexible Electronics. *Adv. Mater.,* **2007**, *19*(15), 1897-1916.
[http://dx.doi.org/10.1002/adma.200602223]

[28] Das, R.; Panda, S.N. Preparation and applications of biochar based nanocomposite: A review. *J. Anal. Appl. Pyrolysis,* **2022**, *167*, 105691.
[http://dx.doi.org/10.1016/j.jaap.2022.105691]

[29] Das, C.; Tamrakar, S.; Kiziltas, A.; Xie, X. Incorporation of Biochar to Improve Mechanical, Thermal and Electrical Properties of Polymer Composites. *Polymers (Basel),* **2021**, *13*(16), 2663.
[http://dx.doi.org/10.3390/polym13162663] [PMID: 34451201]

[30] Zare, Y.; Rhee, K.Y. A simple methodology to predict the tunneling conductivity of polymer/CNT nanocomposites by the roles of tunneling distance, interphase and CNT waviness. *RSC Advances,* **2017**, *7*(55), 34912-34921.
[http://dx.doi.org/10.1039/C7RA04034B]

[31] Liu, Z.; Peng, W.; Zare, Y.; Hui, D.; Rhee, K.Y. Predicting the electrical conductivity in polymer carbon nanotube nanocomposites based on the volume fractions and resistances of the nanoparticle, interphase, and tunneling regions in conductive networks. *RSC Advances,* **2018**, *8*(34), 19001-19010.
[http://dx.doi.org/10.1039/C8RA00811F] [PMID: 35539634]

[32] Taherian, R. Experimental and analytical model for the electrical conductivity of polymer-based

nanocomposites. *Compos. Sci. Technol.,* **2016**, *123*, 17-31.
[http://dx.doi.org/10.1016/j.compscitech.2015.11.029]

[33] Psarras, G.C. Hopping conductivity in polymer matrix–metal particles composites. *Compos., Part A Appl. Sci. Manuf.,* **2006**, *37*(10), 1545-1553.
[http://dx.doi.org/10.1016/j.compositesa.2005.11.004]

[34] Zare, Y.; Rhee, K.Y. A power model to predict the electrical conductivity of CNT reinforced nanocomposites by considering interphase, networks and tunneling condition. *Compos., Part B Eng.,* **2018**, *155*, 11-18.
[http://dx.doi.org/10.1016/j.compositesb.2018.08.028]

[35] Vargas-Bernal, R.; Herrera-Pérez, G.; Calixto-Olalde, M.E.; Tecpoyotl-Torres, M. Analysis of DC Electrical Conductivity Models of Carbon Nanotube-Polymer Composites with Potential Application to Nanometric Electronic Devices. *J. Electr. Comput. Eng.,* **2013**, *2013*, 1-14.
[http://dx.doi.org/10.1155/2013/179538]

[36] Li, S.; Li, X.; Deng, Q.; Li, D. Three kinds of charcoal powder reinforced ultra-high molecular weight polyethylene composites with excellent mechanical and electrical properties. *Mater. Des.,* **2015**, *85*, 54-59.
[http://dx.doi.org/10.1016/j.matdes.2015.06.163]

[37] Giorcelli, M.; Savi, P.; Khan, A.; Tagliaferro, A. Analysis of biochar with different pyrolysis temperatures used as filler in epoxy resin composites. *Biomass Bioenergy,* **2019**, *122*, 466-471.
[http://dx.doi.org/10.1016/j.biombioe.2019.01.007]

[38] Gabhi, R.S.; Kirk, D.W.; Jia, C.Q. Preliminary investigation of electrical conductivity of monolithic biochar. *Carbon,* **2017**, *116*, 435-442.
[http://dx.doi.org/10.1016/j.carbon.2017.01.069]

[39] Kane, S.; Ulrich, R.; Harrington, A.; Stadie, N.P.; Ryan, C. Physical and chemical mechanisms that influence the electrical conductivity of lignin-derived biochar. *Carbon Trends,* **2021**, *5*, 100088.
[http://dx.doi.org/10.1016/j.cartre.2021.100088]

[40] Rehrah, D.; Reddy, M.R.; Novak, J.M.; Bansode, R.R.; Schimmel, K.A.; Yu, J.; Watts, D.W.; Ahmedna, M. Production and characterization of biochars from agricultural by-products for use in soil quality enhancement. *J. Anal. Appl. Pyrolysis,* **2014**, *108*, 301-309.
[http://dx.doi.org/10.1016/j.jaap.2014.03.008]

[41] Nan, N.; DeVallance, D.B.; Xie, X.; Wang, J. The effect of bio-carbon addition on the electrical, mechanical, and thermal properties of polyvinyl alcohol/biochar composites. *J. Compos. Mater.,* **2016**, *50*(9), 1161-1168.
[http://dx.doi.org/10.1177/0021998315589770]

[42] Poulose, A.M.; Elnour, A.Y.; Anis, A.; Shaikh, H.; Al-Zahrani, S.M.; George, J.; Al-Wabel, M.I.; Usman, A.R.; Ok, Y.S.; Tsang, D.C.W.; Sarmah, A.K. Date palm biochar-polymer composites: An investigation of electrical, mechanical, thermal and rheological characteristics. *Sci. Total Environ.,* **2018**, *619-620*, 311-318.
[http://dx.doi.org/10.1016/j.scitotenv.2017.11.076] [PMID: 29154049]

[43] Li, S., Huang, A., Chen, Y.-J., Li, D., Turng, L.-S. Highly filled biochar/ultra-high molecular weight polyethylene/linear low density polyethylene composites for high-performance electromagnetic interference shielding. *Composites Part B: Engineering,* **2018**, *153*, 277-284.
[http://dx.doi.org/10.1016/j.compositesb.2018.07.049] [PMID: 29154049]

CHAPTER 8

Coconut Shell Derived Carbon Reinforced Polymer Composite Films for Packaging Applications

Gautam Chandrasekhar[1] and **Vijaya Rangari**[1,*]

[1] *Department of Materials Science and Engineering, Tuskegee University, Tuskegee, AL 36088, USA*

Abstract: With the advancement toward global sustainability, there is a widespread demand for sustainable materials that can be used for various applications. Carbon has gained much attention in the past few decades due to its scope of utilization in energy and environment related applications. Biomass resources are considered a prominent precursor for the synthesis of carbon-based materials due to their availability and economic viability. In this study, high-quality graphitic carbon is synthesized from Coconut Shell Powder (CSP) by pyrolysis and reinforced into a low-density polyethylene (LDPE) matrix for fabricating films for packaging applications. A custom-built high-temperature autogenic pressure reactor was used for conducting the pyrolysis to synthesize carbon from the coconut shell powder and a blown film extruder was used for fabricating composite films. For preparing the films, coconut shell powder-derived carbon was added to the LDPE matrix at various weight percent loadings of 0.25, 0.5, and 1 wt.%, respectively. Various analytical techniques such as scanning electron microscopy, X-ray diffraction, Raman spectroscopy, thermogravimetric analysis, tensile test, and differential scanning calorimetry were used for studying the properties of carbon and LDPE/carbon composite films. Upon adding carbon as fillers, there were significant improvements in the tensile and thermal degradation properties of the polymer carbon composite films. Upon the incorporation of carbon into the LDPE matrix, the crystallinity and tensile strength were found to improve by a maximum of 29% and 13%, respectively.

Keywords: Biochar, Blow Films, Coconut Shell Powder, Film Extrusion, Food Packaging, LDPE, Low-density polyethylene, Raman, SEM, XRD.

1. INTRODUCTION

The global demand for materials that are more sustainable, efficient, and economical is rising exponentially along with the population. Due to this, the

* **Corresponding author Vijaya Rangari:** Department of Materials Science and Engineering, Tuskegee University, Tuskegee, AL 36088, USA; E-mail: vrangari@tuskegee.edu

Deepa Kodali & Vijaya Rangari (Eds.)

development of sustainable materials has become one of the most prominent needs of society. Carbon-based materials (CBM) are the most versatile materials used for energy and environmental applications [1, 2]. In earlier days, CBM with properties such as optimum surface area and porosity were widely synthesized from fossil-derived precursors using complex techniques such as the electric discharge method and chemical vapor deposition. The most popular synthetic route to prepare CBM with high-surface-area and porosity is the nano-casting approach but this also involves several limitations of being high cost and complicated. Therefore, the development of novel and improvised approaches that are easily scalable, economical, and efficient for the synthesis of CBM from renewable natural precursors is highly desired. Among the CBM synthesized from biomass, different forms of carbon such as carbon nanotubes, graphene, graphene oxide, reduced graphene oxide, activated carbon, *etc.* have gained popularity [3 - 5].

There are widespread carbon-based natural resources created by nature, coupling carbon with other elements. Nowadays, these biomass renewable resources are being widely used as precursors for the synthesis of CBM and they aid in the sustainable development of human society. Some of the common precursors for obtaining CBM are plants [6 - 8], aquatic organisms [9, 10], and animals [11]. Several thermal methods are being used for the synthesis of CBM from biomass such as pyrolysis, hydrothermal carbonization, microwave irradiation, and laser processing. The properties of the CBM highly depend on the composition of the precursor [5].

1.1. Overview of Plant-derived Carbon-based Materials

Plant-based CBMs are commonly derived from plants and their byproducts which are lignocellulose rich materials. Plant-based biomass resources can be commonly classified into two categories which are woody and non-woody resources. Woody resources include materials composed mainly of lignin, cellulose, and hemicellulose whereas non-woody resources consist of materials that are composed of additional components such as lipids, proteins, sugar, starch, *etc* [12]. The carbon derived from bio-based resources is often known as biochar. Some of the common lignocellulose precursors for CBM are palm shells, wood birch, olive waste, wheat straw, hemp, coir, jute, abaca, flax and coconut shell. The composition of precursors has an important role in the properties of the CBM derived from them. In former studies, precursors having a high lignin content have been found to be ideal for getting CBM with high porosity and surface area [13]. Coconut shells are widely used precursors for synthesizing CBM. The composition of lignin, cellulose, and hemicellulose is 46%, 14%, and 32%, respectively [14]. Due to the high content of lignin, coconut shell-derived carbon is desired to have high porosity [13]. Coconut shell-derived CBMs are often used for applications such as reinforcement fillers [15], dye removal [16], supercapacitor [17], and water filtration [18].

1.2. Biochar Carbon-reinforced Polymer Composites

Biochar, a renewable carbonaceous material synthesized through thermochemical processes conducted in an oxygen-restricted environment is an alternate environment-friendly option to reduce the dependency on fossil fuel-derived carbon materials. Since biochar is obtained from sustainable bio-based resources and possesses supreme properties such as high surface area, high thermal stability, and electrical conductivity, it has gained great attention recently among both academic and industrial researchers [19]. The process of carbonization of raw material in an oxygen-free environment is called pyrolysis. It is one of the most common methods preferred for the synthesis of biochar among numerous methods like combustion, liquefication, gasification, *etc.* The properties of biochar obtained from pyrolysis depend on various process parameters such as heating rate, temperature, the type of raw material, and residence time. Lignocellulose rich precursors are often considered as ideal raw materials for synthesizing high-quality biochar. Along with biochar, pyrolysis of biomass can also give some valuable byproducts such as bio-oil and other useful chemicals [20]. Reinforcement of biochar in a polymer matrix is found to be very efficient as it helps in the enhancement of the properties of the polymer. In a work done by Poulose *et al.*, reinforcement of biochar synthesized from pyrolysis of date palm waste into a polypropylene matrix was found to improve the electrical and mechanical properties of the polypropylene matrix considerably [21]. In another study, biochar obtained from the pyrolysis of starch-based packaging waste was modified using ultrasonication and was added to a polypropylene matrix for effective reinforcement and it was found that the modified biochar helped considerably improving the mechanical and thermal properties [22]. Coconut shell, a bio-based waste present abundantly in Asian and South American countries is considered a suitable precursor for pyrolysis to synthesize biochar because of its high lignin content. Researchers have used coconut shell-derived biochar as fillers in various polymers. When coconut shell-derived biochar was added to a polylactic acid/ polybutylene adipate-co-terephthalate matrix, the mechanical and electrical properties were found to improve as a result of superior interfacial interaction between the polymer blend matrix and biochar [23]. Similarly, there are numerous studies portraying the improvement of polymer matrix properties upon the incorporation of coconut shell-derived carbon fillers [7, 24].

Low-density polyethylene (LDPE) is a widely used polymer system for packaging applications because of its flexibility, strength, and durability. It has been found that the addition of fillers into the LDPE matrix is effective in improving their properties [25]. Since it is evident that coconut shell-derived carbon enhances the properties of polymers when used as fillers. In this study, carbon synthesized from coconut shells by means of pyrolysis is reinforced into an LDPE matrix to fabricate composite films for packaging applications.

2. MATERIALS AND METHODS

2.1. Materials

LDPE pellets were purchased from Premier Polymers LLC (EXXONMOBIL ENABLE 2010M).

2.2. Synthesis of Carbon by Pyrolysis

Coconut shell powder was subjected to autogenic pyrolysis in MTI Inc. autogenic pressure reactor. The heating rate was 5° C/minute, up to 850° C. The sample was then kept isothermally at 850°C for 150 minutes for promoting effective graphitization. The coconut shell powder carbon (CSPC) obtained was then ground into a fine powder by using mortar and pastel.

2.3. Characterization of the Synthesized Carbon

2.3.1. X-ray Diffraction (XRD)

Rigaku Smart lab II X-Ray Diffractometer with Cu-Kα radiation ($\lambda = 1.54$ Å) was used for the X-ray diffraction analysis of CSPC. The analysis was carried out at a scan speed of 5°C/min over a range of 10° to 80°.

2.3.2. Raman Spectroscopy

Raman spectroscopy was conducted using a Thermo-Scientific DXR Raman spectroscopy with a laser wavelength of 780 nm. The laser power was 5.0 mW and the analysis was done within the spectrum range of 500 to 3500 cm^{-1}.

2.3.3. Scanning Electron Microscopy (SEM)

JEOL JSM 7200F SEM was used to analyze the morphology of synthesized coconut shell powder carbon (CSPC) and the fracture surface of the films.

2.4. Blown Film Extrusion

The films were synthesized using a LABTECH ultra micro blown film extrusion machine. The CSPC was added to the LDPE pellets at various loadings of 0, 0.25, 0.5, and 1 wt.% and was mixed well until the pellets were uniformly coated with carbon. The different mixtures of CSPC and LDPE pellets were then subjected to blow film extrusion. The screw temperature and the die temperature were set at 140° C and 120° C, respectively. The fabricated films had an average thickness of 0.05 mm to 0.07 mm.

2.5. Analysis of Films

2.5.1. Thermogravimetric Analysis

Thermogravimetric analysis of CSPC/LDPE films was carried out using a TA Q500 TGA analyzer. The samples of approximately 14 mg each were heated from 30°C to 700°C at a heating rate of 10°C/minute in the presence of nitrogen gas.

2.5.2. Differential Scanning Calorimetry

A TA Q2000 Differential Scanning Calorimetry (DSC) was utilized to study the thermal behaviors of the CSPC/LDPE composite films. The samples were 9 to 12 mg each and were heated at a rate of 5° C/minute from 30°C to 250°C in the presence of nitrogen gas. The degree of crystallinity was also determined for each sample using the total enthalpy method which is based on the following equation.

$$Xc = \frac{\Delta Hm}{(1 - \varphi)\Delta H^\circ m} \times 100$$

Where Xc is the degree of crystallinity, ΔHm is the melting enthalpy of the sample, $\Delta H^\circ m$ is the melting enthalpy of 100% crystalline polyethylene (293 J/g) and φ is the weight fraction of CSPC added to the matrix [26].

2.5.3. Tensile Test

Tensile tests of the CSPC/LDPE composite films were conducted using a Zwick Roel Z2.5 Universal Mechanical Testing Machine with a load cell of 2.5 kN load cell. The tests were conducted as per the ASTM standards.

3. RESULTS

3.1. Scanning Electron Microscopy of CSPC

Fig. (**1**) represents the SEM micrographs of CSPC. The surface morphology evidently shows the layer-like arrangement of disordered graphitic planes. The flaky and agglomerated stacked structure also indicates the presence of reduced graphene oxide [27].

Fig. (1). SEM micrograph of CSPC.

3.2. X-ray Diffraction

In Fig. (**2**), the XRD analysis of CSPC shows two peaks at 24.3° and 44.17° which represent the (002) plane and (001) plane respectively. The XRD spectrum of CSPC strongly suggests the presence of reduced graphene oxide. The broad peak of 002 plane can be attributed to the random arrangement of graphitic layers of carbon. Also, the d-spacing corresponding to (002) plane was 3.65 Å which can be attributed to the close stacking of multiples layers of graphene with less oxygen containing functional groups between them [27, 28]. The peak at 44.17° which represents (001) plane can be attributed to the graphene planes present in the synthesized carbon [28].

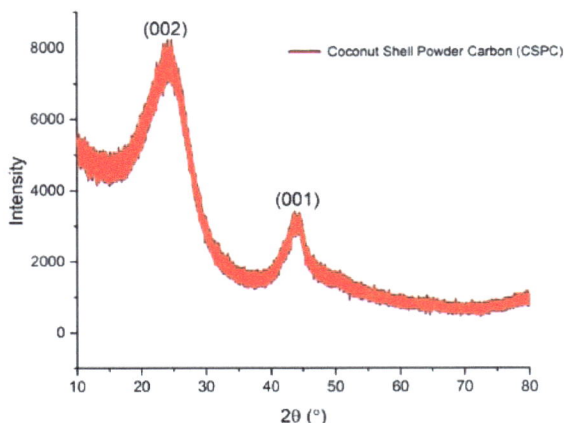

Fig. (2). XRD spectrum of CSPC.

3.3. Raman Spectroscopy

Fig. (**3**) portrays the Raman spectrum of CSPC. D and G bands were observed at 1317.83 cm^{-1} and 1588.80 cm^{-1} respectively. G band resulting from the first-order scattering of the E_{2g} phonon from sp2 carbon atoms present in graphitic carbon confirms the graphitic nature of CSPC. D band vibration formed from the breathing mode of j-point photons of A_{1g} symmetry can also be seen in the spectrum. The defects in the sample are relatively high which can be reasoned to the high-temperature pyrolysis process. The I_D/I_G ratio was found to be 1.16. Broad 2D bands were also present at around 2720.8 cm^{-1} and 2901.1 cm^{-1} which confirms the presence and formation of multi-layered reduced graphene oxide in CSPC [29 - 31].

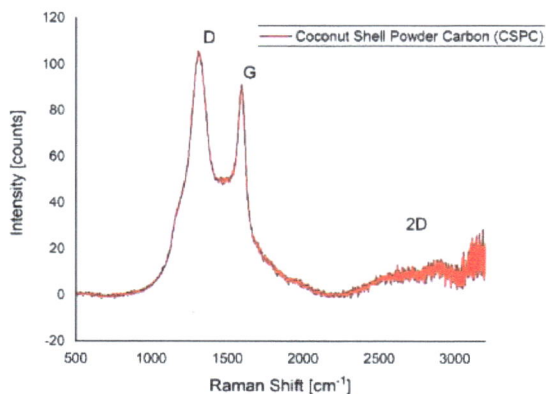

Fig. (3). Raman spectrum of CSPC.

3.4. Thermogravimetric Analysis

Thermal degradation behavior was analyzed using TGA and is represented in Fig. (4) and Table (1). Upon adding CSPC to the LDPE matrix, the onset of degradation was found to improve considerably by about a maximum of 10° Celsius. This can be attributed to the effective crosslinking and interfacial adhesion between the CSPC and LDPE matrix. The composite film with 0.5wt. % of CSPC was found to have the highest onset of degradation temperature among all the samples. This increment in thermal stability makes the composite films suitable for relatively higher temperature applications compared to the neat LDPE films.

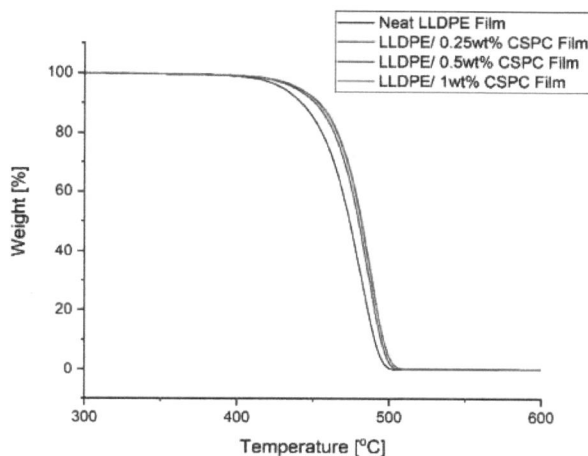

Fig. (4).   Thermal analysis of LDPE/CSPC films; TGA weight% change with temperature.

Table 1. TGA results of LDPE/CSPC films.

Sample	Onset Decomposition Temperature (°C)
Neat LDPE film	455.13
LDPE/0.25 wt.% CSPC film	463.13
LDPE/0.5 wt.% CSPC film	466.41
LDPE/1 wt.% CSPC film	465.29

3.5. Differential Scanning Calorimetry

DSC analysis was also incorporated for studying the thermal behavior of the films which is shown in Fig. (5) and Table (2). The melting point of the CSPC/LDPE composite films was found to decrease by a small amount compared to the neat LDPE film. However, the crystallinity was found to increase upon the addition of

CSPC fillers. The crystallinity was found to improve by 12-29% with the incorporation of CSPC.

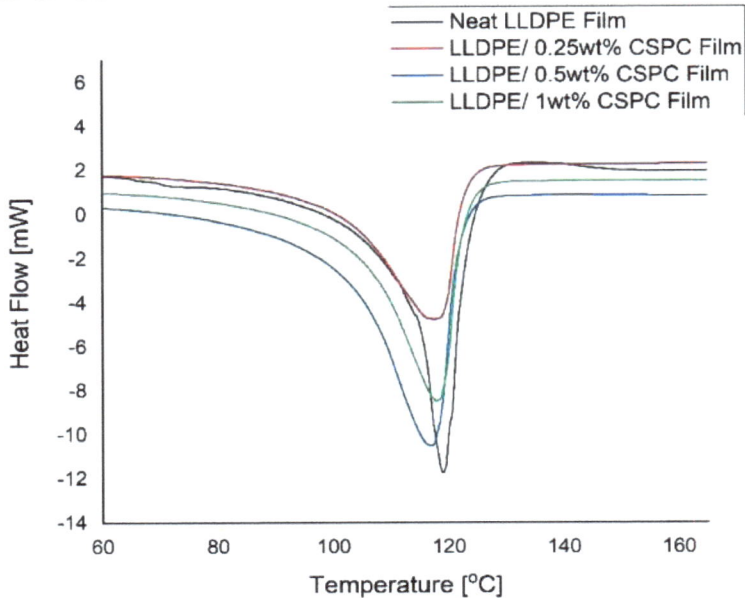

Fig. (5). Thermal analysis of LDPE/CSPC films; DSC thermographs.

Table 2. DSC results of LDPE/CSPC films.

Sample	Melting Temperature, Tm (°C)	Melting Enthalpy (J/g)	Crystallinity%
Neat LDPE film	119.6	89.61	30.58
LDPE/0.25 wt.% CSPC film	117.2	100.4	34.53
LDPE/0.5 wt.% CSPC film	117.05	104.9	35.98
LDPE/1 wt.% CSPC film	117.83	115.2	39.71

3.6. Tensile Test

The mechanical behavior was analyzed by means of tensile testing. The tensile test results are presented in Fig. (**6**) and Table (**3**). Upon the addition of CSPC, both strength and the percentage of elongation were found to increase. 0.25 wt. % CSPC/LDPE films were found to have the highest elongation. However, the elongation was found to decrease with the increment of CSPC in the LDPE matrix. 1 wt.% CSPC/LDPE films were found to have high strength, but the elongation percentage was even lower than the neat LDPE films which can be attributed to the disturbance caused in the matrix due to the agglomeration of CSPC.

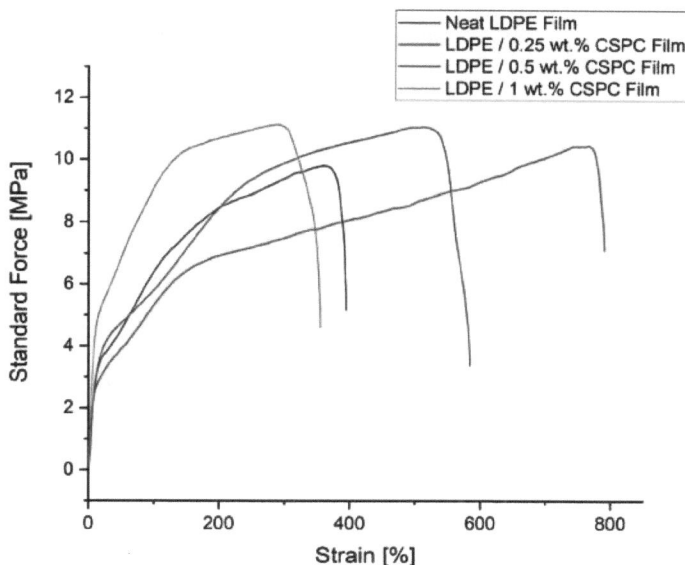

Fig. (6). Tensile test results of LDPE/CSPC composite films.

Table 3. Tensile results of LDPE/CSPC films

Sample	Maximum Strength (MPa)
Neat LDPE film	9.81
LDPE/0.25 wt.% CSPC film	10.47
LDPE/0.5 wt.% CSPC film	11.07
LDPE/1 wt.% CSPC film	11.13

3.7. Fracture Surface Analysis of LDPE/CSPC Films using SEM

The composite film with 1wt.% CSPC was found to have the maximum tensile strength compared to others. Hence, its fracture surface morphology was analyzed in comparison with the neat LDPE films and is shown in Fig. (7). The neat LDPE film fracture surface (Fig. **7(A, C)**) had more pullouts when compared to that of LDPE/1 wt.% CSPC film which can be accounted for its relatively higher strain behavior and absence of reinforcements. Some voids were also observed on the neat film which might have been formed during their fabrication process. In case of LDPE/1 wt.% CSPC film's fracture surface (Fig. **7(B, D)**), considerably low voids and fibrous pullouts were observed. The fracture surface of the composite film evidently shows that the reinforcement was effective and uniform dispersion of CSPC in the LDPE matrix was achieved.

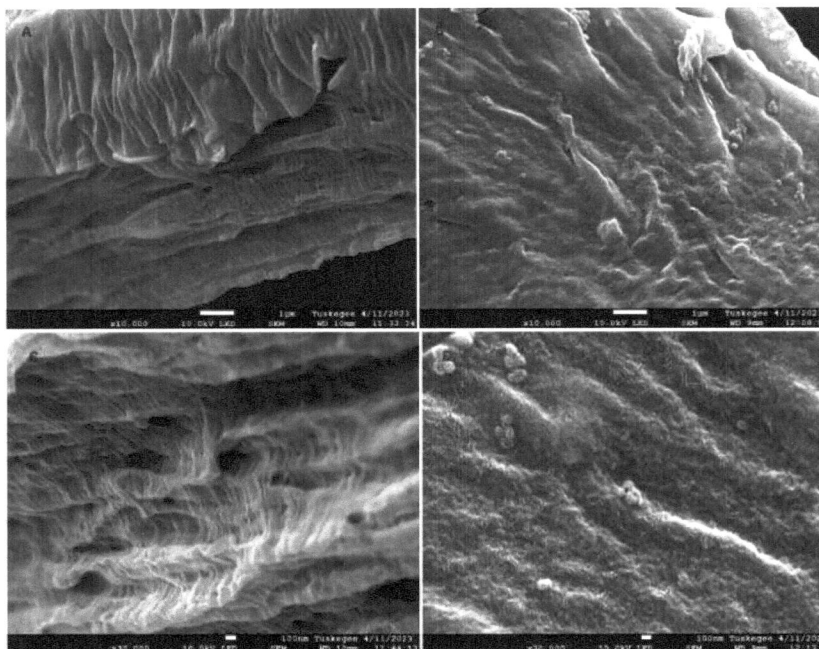

Fig. (7). SEM micrographs of the fracture surface of LDPE/ CSPC films; (**A**, **C**) Neat LDPE film and (**B**, **D**) LDPE/1wt.% CSPC film.

CONCLUSION

Carbon was successfully synthesized from waste coconut shell powder by pyrolysis. XRD, Raman, and SEM analysis revealed the presence of reduced graphene oxide in the synthesized carbon. The synthesized carbon was later successfully incorporated into a low-density polyethylene matrix for fabricating films for packaging applications. Carbon composite films were found to have significantly higher thermal and mechanical properties compared to neat LDPE films. The crystallinity of the matrix was found to improve upon the addition of CSPC by 13% to 29%. Also, the tensile strength was found to improve by 13%. Hence, carbon synthesized from coconut shell powder is highly effective to be used as fillers in LDPE matrix for making composite films which can be used for packaging applications that demand higher mechanical and thermal properties.

FUNDING SOURCES

The authors would like to acknowledge the financial support from AL-EPSCoR #1655280 and NSF CREST #1735971.

REFERENCES

[1]	Titirici, M.M.; Antonietti, M. Chemistry and materials options of sustainable carbon materials made by hydrothermal carbonization. *Chem. Soc. Rev.,* **2010,** *39*(1), 103-116.
[http://dx.doi.org/10.1039/B819318P] [PMID: 20023841]

[2]	Sharma, H. B.; Sarmah, A. K.; Dubey, B. Hydrothermal carbonization of renewable waste biomass for solid biofuel production: A discussion on process mechanism, the influence of process parameters, environmental performance and fuel properties of hydrochar *Renew. sustain. energy rev.,* **2020,** *123*.
[http://dx.doi.org/10.1016/j.rser.2020.109761]

[3]	Khan, T. A.; Saud, A. S.; Jamari, S. S.; Rahim, M. H. A.; Park, J. W.; Kim, H. J. Hydrothermal carbonization of lignocellulosic biomass for carbon rich material preparation: A review *Biomass and Bioenergy,* **2019,** *130*.
[http://dx.doi.org/10.1016/j.biombioe.2019.105384]

[4]	Zhang, M.; He, L.; Shi, T.; Zha, R. Nanocasting and Direct Synthesis Strategies for Mesoporous Carbons as Supercapacitor Electrodes *Chemistry of Materials,* **2018,** *30*(21), 7391-7412.
[http://dx.doi.org/10.1021/acs.chemmater.8b03345]

[5]	Wang, Y. Biomass-Derived Carbon Materials: Controllable Preparation and Versatile Applications *Small,* **2021,** *17*(40).
[http://dx.doi.org/10.1002/smll.202008079]

[6]	Pitchaiya, S.; Eswaramoorthy, N.; Natarajan, M.; Santhanam, A.; Asokan, V.; Madurai Ramakrishnan, V.; Rangasamy, B.; Sundaram, S.; Ravirajan, P.; Velauthapillai, D. Perovskite Solar Cells: A Porous Graphitic Carbon based Hole Transporter/Counter Electrode Material Extracted from an Invasive Plant Species Eichhornia Crassipes. *Sci. Rep.,* **2020,** *10*(1), 6835.
[http://dx.doi.org/10.1038/s41598-020-62900-4] [PMID: 32321928]

[7]	Umerah, C.O.; Kodali, D.; Head, S.; Jeelani, S.; Rangari, V.K. Synthesis of carbon from waste coconutshell and their application as filler in bioplast polymer filaments for 3d printing. *Compos., Part B Eng.,* **2020,** *202*, 108428.
[http://dx.doi.org/10.1016/j.compositesb.2020.108428]

[8]	Dutta, V. Bio-Inspired Synthesis of Carbon-Based Nanomaterials and Their Potential Environmental Applications: A State-of-the-Art Review *Inorganics,* **2022,** *10*, 10.
[http://dx.doi.org/10.3390/inorganics10100169]

[9]	Ahmed, M.J.; Islam, M.A.; Asif, M.; Hameed, B.H. Human hair-derived high surface area porous carbon material for the adsorption isotherm and kinetics of tetracycline antibiotics. *Bioresour. Technol.,* **2017,** *243*, 778-784.
[http://dx.doi.org/10.1016/j.biortech.2017.06.174] [PMID: 28711807]

[10]	Zhou, X.; Jia, Z.; Feng, A.; Wang, X.; Liu, J.; Zhang, M.; Cao, H.; Wu, G. Synthesis of fish skin-derived 3D carbon foams with broadened bandwidth and excellent electromagnetic wave absorption performance. *Carbon,* **2019,** *152*, 827-836.
[http://dx.doi.org/10.1016/j.carbon.2019.06.080]

[11]	Huang, W.; Zhang, H.; Huang, Y.; Wang, W.; Wei, S. Hierarchical porous carbon obtained from animal bone and evaluation in electric double-layer capacitors. *Carbon,* **2011,** *49*(3), 838-843.
[http://dx.doi.org/10.1016/j.carbon.2010.10.025]

[12]	Yahya, M.A. A brief review on activated carbon derived from agriculture by-product, **2018,** *1972*(1).
[http://dx.doi.org/10.1063/1.5041244]

[13]	Zaman Chowdhury, Z. Biomass carbon sorbents. **2013.**

[14]	Sekhon, S. S.; Kaur, P.; Park, J. S. From coconut shell biomass to oxygen reduction reaction catalyst: Tuning porosity and nitrogen doping *Renew. sustain. energy rev.,* **2021,** *147*, 111173.
[http://dx.doi.org/10.1016/j.rser.2021.111173]

[15] Zhang, Q.; Zhang, D.; Lu, W.; Khan, M.U.; Xu, H.; Yi, W.; Lei, H.; Huo, E.; Qian, M.; Zhao, Y.; Zou, R. Production of high-density polyethylene biocomposites from rice husk biochar: Effects of varying pyrolysis temperature. *Sci. Total Environ.,* **2020**, *738*, 139910.
[http://dx.doi.org/10.1016/j.scitotenv.2020.139910] [PMID: 32531606]

[16] Widiyastuti, W.; Fahrudin Rois, M.; Suari, N.M.I.P.; Setyawan, H. Activated carbon nanofibers derived from coconut shell charcoal for dye removal application. *Adv. Powder Technol.,* **2020**, *31*(8), 3267-3273.
[http://dx.doi.org/10.1016/j.apt.2020.06.012]

[17] Mi, J.; Wang, X.R.; Fan, R.J.; Qu, W.H.; Li, W.C. Coconut-shell-based porous carbons with a tunable micro/mesopore ratio for high-performance supercapacitors. In: *Energy & Fuels*; **2012**; pp. 5321-5329.
[http://dx.doi.org/10.1021/ef3009234]

[18] Jibril, M.; Noraini, J.; Poh, L.S.; Evuti, A.M. *Jurnal Teknologi Removal of Colour from Waste Water Using Coconut Shell Activated Carbon (CSAC) and Commercial Activated Carbon*; CAC, **2013**, pp. 2180-3722. Available from : www.jurnalteknologi.utm.my

[19] Bartoli, M.; Arrigo, R.; Malucelli, G.; Tagliaferro, A.; Duraccio, D. Recent Advances in Biochar Polymer Composites *Polymers,* **2022**, *14*(12), 2506.
[http://dx.doi.org/10.3390/polym14122506]

[20] Das, C.; Tamrakar, S.; Kiziltas, A.; Xie, X. Incorporation of biochar to improve mechanical, thermal and electrical properties of polymer composites *Polymers,* **2021**, *13*, 16.
[http://dx.doi.org/10.3390/polym13162663]

[21] Poulose, A.M.; Elnour, A.Y.; Anis, A.; Shaikh, H.; Al-Zahrani, S.M.; George, J.; Al-Wabel, M.I.; Usman, A.R.; Ok, Y.S.; Tsang, D.C.W.; Sarmah, A.K. Date palm biochar-polymer composites: An investigation of electrical, mechanical, thermal and rheological characteristics. *Sci. Total Environ.,* **2018**, *619-620*, 311-318.
[http://dx.doi.org/10.1016/j.scitotenv.2017.11.076] [PMID: 29154049]

[22] Mohammed, Z.; Jeelani, S.; Rangari, V. Effective reinforcement of engineered sustainable biochar carbon for 3D printed polypropylene biocomposites. *Composites Part C: Open Access,* **2022**, *7*, 100221.
[http://dx.doi.org/10.1016/j.jcomc.2021.100221]

[23] George, J.; Jung, D.; Bhattacharyya, D. Improvement of Electrical and Mechanical Properties of PLA/PBAT Composites Using Coconut Shell Biochar for Antistatic Applications. *Appl. Sci. (Basel),* **2023**, *13*(2), 902.
[http://dx.doi.org/10.3390/app13020902]

[24] Ramaraj, B.; Poomalai, P. Ecofriendly poly(vinyl alcohol) and coconut shell powder composite films: Physico-mechanical, thermal properties, and swelling studies. *J. Appl. Polym. Sci.,* **2006**, *102*(4), 3862-3867.
[http://dx.doi.org/10.1002/app.23913]

[25] Silva-Leyton, R.; Quijada, R.; Bastías, R.; Zamora, N.; Olate-Moya, F.; Palza, H. Polyethylene/graphene oxide composites toward multifunctional active packaging films. *Compos. Sci. Technol.,* **2019**, *184*, 107888.
[http://dx.doi.org/10.1016/j.compscitech.2019.107888]

[26] Mohammed, Z.; Jeelani, S.; Rangari, V.K. Effect of Low-Temperature Plasma Treatment on Starch-Based Biochar and Its Reinforcement for Three-Dimensional Printed Polypropylene Biocomposites. *ACS Omega,* **2022**, *7*(44), 39636-39647.
[http://dx.doi.org/10.1021/acsomega.2c02372] [PMID: 36385856]

[27] Tai, M.J.Y. Green synthesis of reduced graphene oxide using green tea extract In: *AIP Conference Proceedings*; American Institute of Physics Inc, **2018**; *2045*(1), 020032.
[http://dx.doi.org/10.1063/1.5080845]

[28] Thakur, S.; Karak, N. Green reduction of graphene oxide by aqueous phytoextracts. *Carbon,* **2012**, *50*(14), 5331-5339.
[http://dx.doi.org/10.1016/j.carbon.2012.07.023]

[29] Hidayah, N.M.S. Comparison on graphite, graphene oxide and reduced graphene oxide: Synthesis and characterization In: *AIP Conference Proceedings*; American Institute of Physics Inc., **2017**, *1892*(1), 150002.
[http://dx.doi.org/10.1063/1.5005764]

[30] Ossonon, B.D.; Bélanger, D. Synthesis and characterization of sulfophenyl-functionalized reduced graphene oxide sheets. *RSC Advances,* **2017**, *7*(44), 27224-27234.
[http://dx.doi.org/10.1039/C6RA28311J]

[31] Ning, F.; Cong, W.; Hu, Y.; Wang, H. Additive manufacturing of carbon fiber-reinforced plastic composites using fused deposition modeling: Effects of process parameters on tensile properties. *J. Compos. Mater.,* **2017**, *51*(4), 451-462.
[http://dx.doi.org/10.1177/0021998316646169]

CHAPTER 9

Carbon Based Polymer Composites in Water Treatment and Filtration

Sabina Yeasmin[1,*] and **Soma Bose**[1]

[1] *Department of Microbiology, Calcutta National Medical College & Hospital, West Bengal, India*

Abstract: The world at large has acknowledged the importance of environmental issues. The depletion of natural resources, such as drinking water, and the emission of greenhouse gases that result in climatic change and the deterioration of human health, are the primary concerns. As urban areas expand rapidly, they exert enormous strain on nearby water supplies, leading to a global freshwater demand surge that is outpacing population expansion. The development of polymer nanocomposites has contributed significantly to the search for viable answers to pressing ecological concerns. Their ability to eradicate pollutants, including gas emissions, heavy metals, and dyes in wastewater has garnered researchers' attention. In this overview, polymer nanocomposites, as well as the composites reinforced with biocarbon that are used in environmentally friendly ways, are discussed in detail. The adsorption mechanism and applications of polymer nanocomposites for the removal of hazardous metal ions and dyes were also studied.

Keywords: Adsorption, Biochar, Biocarbon, Inorganic Pollutants, Metal Ions, Organic Pollutants, Polymer Membranes, Polymer Composites, Sorption, Water Treatment, Water Filtration, Wastewater Treatment.

1. INTRODUCTION

Polymer nanocomposites reinforced with nanofillers grew rapidly and formed novel materials with enhanced properties. They have versatile applications in aerospace, mechanical, electrical [1], catalyst [2], food packaging [3], medicinal [4], and environmental [5, 6] industries. The chemical compatibility of polymer nanocomposites depends on the shape and size of the nanoparticles [7]. If nanomaterials are properly functionalized, they can improve polymer-filler compatibility and prevent agglomeration effects. Different nanoparticles can

---

* **Corresponding author Sabina Yeasmin:** Department of Microbiology, Calcutta National Medical College & Hospital, West Bengal, India; E-mail: sabina.kolkata21@gmail.com

change the stability and degradation of a polymer matrix [8, 9]. Polymer nanoparticles can also help in the crystallization of the polymer matrix [10].

Different inorganic and organic materials are used for the synthesis of membranes that are used for wastewater treatment [11]. In comparison to typical membranes, nanocomposite membranes display improved performance [12]. Ceramics, metals, and glass were categorized as inorganic materials, while mixed nanocomposites and polymeric materials were categorized as organic materials [13]. Inorganic membrane synthesis is an expensive synthesis process due to excessive chemical resistance, making it suitable for wastewater treatment [14]. Polymeric membranes were a better choice than inorganic membranes in industrial applications. Owing to their versatile applications and simple preparation methods, polymeric membranes were chosen over inorganic membranes [15]. For water decontamination, excellent permeability, retention, and outstanding antifouling properties of membranes were quintessential [16].

Out of all separation technologies available, adsorption is the most efficient, cost-effective, and easy-to-use wastewater treatment technology for contaminant separation [17]. In recent years, biocarbon has emerged as a new sorbent due to its excellent properties, such as being eco-friendly, containing substantial functional groups, porous structures, and having outstanding adsorption capability. These properties have led to biocarbon's widespread use in the process of removing contaminants from wastewater [18].

Waste biomass is used to produce biochar. This is economical and simultaneously beneficial. Biochar that originates from animals or plants has great potential for waste management and can help decrease the pollution. Different variations in waste biomass can be found in nature. It includes crop residues, forestry waste, animal manure, food processing waste, paper mill waste, municipal solid waste and sewage sludge [19]. Animal manure and sewage sludge-based biochar from pyrolysis can kill bacterial populations [20]. Toxic heavy metals are collected from sewage sludge and can have long-term applications. In this book chapter, we discuss the preparation and application of different polymer nanocomposites, along with their antimicrobial properties and their use in water filtration.

2. BIOCHAR IN WATER TREATMENT

The expansion of industry has led to an increase in the amount of waste material disposed of in bodies of water, which in turn has led to contamination of the water. Cd^{2+}, Cu^{2+}, Pb^{2+}, Hg^{2+}, Ni^{2+}, and Cr^{6+} are the heavy metals that cause environmental pollution. Heavy metals and organic compounds are removed from industrial wastewater by using biochar. Biochar, also known as pyrogenic black carbon, is created when organic matter like wood or grass is heated in the absence of nitrogen or oxygen. Biochar has been proposed as a cost-effective adsorbent for use in water treatment because of its unique surface qualities. If chitosan is added on the surface of biochar, it can remove three heavy metals like Cd^{2+}, Pb^{2+} and Cu^{2+} [21]. Zhou *et al.* demonstrated that applying a chitosan coating to biochar surfaces improved its adsorption and soil amendment properties. Chitosan-modified biochars were more effective in removing Pb, Cu, and Cd from the solution. Lead sorption on chitosan-modified biochar lowered metal toxicity as well [22]. The biochar that is obtained from the pyrolysis of wheat straw is used to remove Cd^{2+}, Pb^{2+} and Cu^{2+} from aqueous solutions [23]. Pyrolysis of malt wasted rootlets at 850 °C yielded biochar with 63% micropore volume. Malt-based biochar removed Hg(II) from pure aqueous solutions and has shown good sorption and kinetics. It may adsorb 103 mg/g at pH 5. The sorption kinetics were represented by a pseudo second-order model, and the Langmuir isotherm model was shown to be a good fit for the experimental data [24]. $ZnCl_2$-modified biochar from glue residue was created for the sorption of Cr(VI), which showed a maximum capacity of 325.5 mg/g compared to other sorbents. The results demonstrated that the modified biochar has the largest specific surface area and the most functional groups on the surface, making it an effective and reusable adsorbent for Cr(VI) [25].

Wastage from textile industry also causes water pollution. Bamboo powder chemically activated with phosphoric acid at 400–600°C yields nanoporous carbon with high surface area (NCMs), which can absorb Lanosyn orange and Lanosyn gray effectively [26]. This research showed that high surface area NCMs made from agricultural byproducts have the potential to serve as efficient and inexpensive adsorbent materials for the treatment of dye-contaminated industrial effluent.

The presence of new organic pollutants in industrial effluent, like phenols and polycyclic aromatic hydrocarbons (PAHs), has raised significant alarm. The biochar from sewage sludge was pyrolyzed at 500 °C in the presence of strong HCl. Activated carbons (ACs) made from sludge were made using both traditional heating methods and microwave pyrolysis. The ACs were employed to remove six phenolic compounds from aqueous solutions, and their performance was assessed using a number of analytical and functional methods. All of the adsorbents shared the same characteristics and were hydrophobic on the outside. The mesoporous materials used for the ACs have specific surface areas of up to 641 m2 g1 for the CAC-500 and 540 m2 g1 for the MAC-980. This study proposes that π–π interactions are key to adsorption [27]. At 800°C, biochar from malt spent rootlets can adsorb phenanthrene (PHE). It is higher in magnitude in comparison to other raw materials [28]. Biochar from orange peels was derived from pyrolysis at temperatures ranging from 150 to 700 °C (OP 150-OP 700).

Water molecules increase the polar surface area at lower temperatures along with the interactions with 1-naphthol and show higher sorption capacity than naphthalene (NAP) [29].

Pesticides are toxic and destroy the ecological balance and the environment. Biochar is a promising agent in the removal of pesticides from the environment. Thiacloprid pesticide is removed by the use of biochar prepared from the straw of maize [30]. Almond shell biochar also prepared by pyrolysis helps in the removal of pesticides [31]. Antibiotics dispersed in the natural water are very difficult to decompose. Antibiotic toxicity can also be removed by biochar. The two most common antibiotics Tetracyclines (TCs) and sulphonamides (SAs), are used to kill bacteria and become hazardous for the environment [32]. Biochar prepared from sawdust removes $ZnCl_2/FeCl_3$ solution. Biochar can remove TCs from water after 3 cycles. Sulphonamides adsorb on the surface of the biochar after 3 cycles [33].

Urban storm water contains indicator organisms and pathogens with a lot of contaminants which ultimately integrates with underground or surface water. Vegetables mixed with this water produce a lot of bacteria. Rice husk and rice husk derived biochar were used as filters to treat wastewater. Negatively charged bacteria and pathogens were removed from the environment by the biochar filtration method [34, 35].

Biochar can remove inorganic ions, such as N, P and F from wastewater as inorganic waste. NH_4^+ was adsorbed by hydrous bamboo biochar [36]. Walnut shell and sewage sludge were prepared by co-pyrolysis and removed phosphate ions from eutrophic water [37]. Phosphate ions can be removed from water by biochar by magnetic method [38].

3. POLYMER NANOCOMPOSITES IN WATER FILTRATION

Nanocomposite membranes were synthesized to improve various polymeric membranes. Nanoparticles are incorporated in polymeric membranes, resulting in the outcome of polymeric nanocomposite membranes [11, 39]. Nanocomposite and thin-film nanocomposite membranes were synthesized using two different processes. In the first method, nanoparticles are deposited on the surface of the polymeric membrane. In the second method, mixed matrix nanocomposite membranes were synthesized [40]. Coating/deposition and mixing were some of the various processes used to prepare multifunctional nanocomposite membranes [11]. Chemical groups were generally attached to the surface stratum of the membrane. Additionally, different engineered nanoscale materials were also attached to the polymeric membranes. Carbon nanotubes attached to the polymeric membranes exhibited excellent thermal, electrical and mechanical properties [41]. Carbon nanotubes can also change the physical and chemical properties of polymeric membranes [42]. Due to the presence of interior pores, carbon nanomaterial composite membranes showed enhanced penetrability [43]. Carbon nanotube composite membranes were used to vary temperature and operation [44]. When polymeric membranes are combined with carbon nanotubes, there is an increase in each and every property of the combined product.

To separate oil from wastewater, a special type of nanotube-polymer composite membrane was synthesized using a polyvinyl alcohol barricade stratum [45]. The chemical vapour deposition technique was combined with the phase inversion technique for carbon nanotube synthesis. This method was also used to fabricate membranes. The polymer composite membrane with 7.5% CNTs has 119% higher tensile strength, 77% higher Young's modulus, and 258% higher toughness than the baseline polymer. The membrane permeate has oil contents below 10 mg/L and oil rejection of over 95%.

Zwitterion-functionalized CNTs in a polyamide membrane enhanced water flux and salt rejection. As the CNT fraction grew from 0 to 20%, water flux increased by more than a factor of 4, from 6.8 to 28.7 GFD (gallons per square foot per day). The ion rejection ratio improved from 97.6% to 98.6%. Thus, nanotubes added a transport mechanism to the polyamide membrane that increased flow rate and selectivity. Simulations reveal that when two zwitterions are coupled to each end of 15-Å CNTs, the ion rejection ratio is nearly 100%. Nonfunctionalized CNTs have a rejection percentage of 0%, while CNTs with five carboxylic acid groups per end have 20% [46]. A new polyphenol-metal influenced nanohybridization method was used to fabricate carbon nanomaterials attached on the membrane. It was also used to remove harmful bacteria in the water [47]. Membrane hydrophobicity is also a great problem like membrane roughness. Different kinds of nanoparticles were incorporated into polymeric membranes. This incorporation decreases water fouling and serves as antibacterial agents [48].

CONCLUSION

In comparison to other materials, polymer nanocomposite, biocarbon, and biochar can easily purify wastewater. Depending on the size and shape, composition, and substitutes added on the surface, they can be easily applied in versatile fields. Biochar can have multiple benefits including climate change mitigation, improving soil fertility, waste management, energy production, contaminant removal, and compost preparation. Polymer nanocomposites are also promising agents in the biomedical field. Carbon nanotubes serve as antifouling agents and are known for their membrane roughness, hydrophobicity, and rough surface properties. When negatively charged hydrophilic groups are added, they become more hydrophilic. Graphene oxide, metallic nanoparticles, and biochar are all used for the treatment of wastewater. Pesticides, antibiotics, indicator organisms and pathogens, and inorganic ions can be removed from the environment with the help of biochar.

REFERENCES

[1] Abdullahi Hassan, Y.; Hu, H. Current status of polymer nanocomposite dielectrics for high-temperature applications. *Compos., Part A Appl. Sci. Manuf.,* **2020**, *138*, 106064.
[http://dx.doi.org/10.1016/j.compositesa.2020.106064]

[2] Ahmad, N.; Anae, J.; Khan, M.Z.; Sabir, S.; Yang, X.J.; Thakur, V.K.; Campo, P.; Coulon, F. Visible light-conducting polymer nanocomposites as efficient photocatalysts for the treatment of organic pollutants in wastewater. *J. Environ. Manage.,* **2021**, *295*, 113362.
[http://dx.doi.org/10.1016/j.jenvman.2021.113362] [PMID: 34346390]

[3] Cheikh, D.; Majdoub, H.; Darder, M. An overview of clay-polymer nanocomposites containing bioactive compounds for food packaging applications. *Appl. Clay Sci.,* **2022**, *216*, 106335.
[http://dx.doi.org/10.1016/j.clay.2021.106335]

[4] Bharadwaz, A.; Jayasuriya, A.C. Recent trends in the application of widely used natural and synthetic polymer nanocomposites in bone tissue regeneration. *Mater. Sci. Eng. C,* **2020**, *110*, 110698.
[http://dx.doi.org/10.1016/j.msec.2020.110698] [PMID: 32204012]

[5] Yashas, S.R.; Shahmoradi, B.; Wantala, K.; Shivaraju, H.P. Potentiality of polymer nanocomposites for sustainable environmental applications: A review of recent advances. *Polymer,* **2021**, *233*, 124184.
[http://dx.doi.org/10.1016/j.polymer.2021.124184]

[6] Momina, ; Ahmad, K. Study of different polymer nanocomposites and their pollutant removal efficiency: Review. *Polymer,* **2021**, *217*, 123453.
[http://dx.doi.org/10.1016/j.polymer.2021.123453]

[7] Muhammed Shameem, M.; Sasikanth, S.M.; Annamalai, R.; Ganapathi Raman, R. A brief review on polymer nanocomposites and its applications. *Mater. Today Proc.,* **2021**, *45*, 2536-2539.
[http://dx.doi.org/10.1016/j.matpr.2020.11.254]

[8] Pandey, J.K.; Raghunatha Reddy, K.; Pratheep Kumar, A.; Singh, R.P. An overview on the degradability of polymer nanocomposites. *Polym. Degrad. Stabil.,* **2005**, *88*(2), 234-250.
[http://dx.doi.org/10.1016/j.polymdegradstab.2004.09.013]

[9] Pielichowska, K. Thermooxidative degradation of polyoxymethylene homo- and copolymer nanocomposites with hydroxyapatite: Kinetic and thermoanalytical study. *Thermochim. Acta,* **2015**, *600*, 7-19.
[http://dx.doi.org/10.1016/j.tca.2014.11.016]

[10] Pielichowska, K.; Dryzek, E.; Olejniczak, Z.; Pamuła, E.; Pagacz, J. A study on the melting and crystallization of polyoxymethylene-copolymer/hydroxyapatite nanocomposites. *Polym. Adv. Technol.,* **2013**, *24*(3), 318-330.
[http://dx.doi.org/10.1002/pat.3086]

[11] Al Aani, S.; Wright, C.J.; Atieh, M.A.; Hilal, N. Engineering nanocomposite membranes: Addressing current challenges and future opportunities. *Desalination,* **2017**, *401*, 1-15.
[http://dx.doi.org/10.1016/j.desal.2016.08.001]

[12] Kango, S.; Kalia, S.; Celli, A.; Njuguna, J.; Habibi, Y.; Kumar, R. Surface modification of inorganic nanoparticles for development of organic–inorganic nanocomposites—A review. *Prog. Polym. Sci.,* **2013**, *38*(8), 1232-1261.
[http://dx.doi.org/10.1016/j.progpolymsci.2013.02.003]

[13]   Ulbricht, M. Advanced functional polymer membranes. *Polymer,* **2006**, *47*(7), 2217-2262.
     [http://dx.doi.org/10.1016/j.polymer.2006.01.084]

[14]   Ng, L.Y.; Mohammad, A.W.; Leo, C.P.; Hilal, N. Polymeric membranes incorporated with metal/metal oxide nanoparticles: A comprehensive review. *Desalination,* **2013**, *308*, 15-33.
     [http://dx.doi.org/10.1016/j.desal.2010.11.033]

[15]   Hilal, N.; Ismail, A.F.; Wright, C. *Membrane Fabrication, First*; CRC Press: Boca Raton, **2015**.
     [http://dx.doi.org/10.1201/b18149]

[16]   Yin, J.; Deng, B. Polymer-matrix nanocomposite membranes for water treatment. *J. Membr. Sci.,* **2015**, *479*, 256-275.
     [http://dx.doi.org/10.1016/j.memsci.2014.11.019]

[17]   Liang, L.; Xi, F.; Tan, W.; Meng, X.; Hu, B.; Wang, X. Review of organic and inorganic pollutants removal by biochar and biochar-based composites. *Biochar,* **2021**, *3*(3), 255-281.
     [http://dx.doi.org/10.1007/s42773-021-00101-6]

[18]   Hu, B.; Ai, Y.; Jin, J.; Hayat, T.; Alsaedi, A.; Zhuang, L.; Wang, X. Efficient elimination of organic and inorganic pollutants by biochar and biochar-based materials. *Biochar,* **2020**, *2*(1), 47-64.
     [http://dx.doi.org/10.1007/s42773-020-00044-4]

[19]   Cantrell, K.B.; Hunt, P.G.; Uchimiya, M.; Novak, J.M.; Ro, K.S. Impact of pyrolysis temperature and manure source on physicochemical characteristics of biochar. *Bioresour. Technol.,* **2012**, *107*, 419-428.
     [http://dx.doi.org/10.1016/j.biortech.2011.11.084] [PMID: 22237173]

[20]   Lehmann, J.; Joseph, S. *Biochar for Environmental Management*; Routledge, **2015**.

[21]   Hussain, A.; Maitra, J.; Khan, K.A. Development of biochar and chitosan blend for heavy metals uptake from synthetic and industrial wastewater. *Appl. Water Sci.,* **2017**, *7*(8), 4525-4537.
     [http://dx.doi.org/10.1007/s13201-017-0604-7]

[22]   Zhou, Y.; Gao, B.; Zimmerman, A.R.; Fang, J.; Sun, Y.; Cao, X. Sorption of heavy metals on chitosan-modified biochars and its biological effects. *Chem. Eng. J.,* **2013**, *231*, 512-518.
     [http://dx.doi.org/10.1016/j.cej.2013.07.036]

[23]   Wang, Y.; Zheng, K.; Jiao, Z.; Zhan, W.; Ge, S.; Ning, S.; Fang, S.; Ruan, X. Simultaneous removal of Cu^{2+}, Cd^{2+} and Pb^{2+} by modified wheat straw biochar from aqueous solution: Preparation, characterization and adsorption mechanism. *Toxics,* **2022**, *10*(6), 1-15.
     [http://dx.doi.org/10.3390/toxics11010001] [PMID: 35736924]

[24]   Boutsika, L.G.; Karapanagioti, H.K.; Manariotis, I.D. Aqueous mercury sorption by biochar from malt spent rootlets. *Water Air Soil Pollut.,* **2014**, *225*(1), 1805.
     [http://dx.doi.org/10.1007/s11270-013-1805-9]

[25]   Shi, Y. High-efficiency removal of Cr(VI) by modified biochar derived from glue residue. *J. Clean. Prod.,* **2020**, *254*
     [http://dx.doi.org/10.1016/j.jclepro.2019.119935]

[26]   Pradhananga, R. Wool carpet dye adsorption on nanoporous carbon materials derived from agro-product. *C,* **2017**, *3*(2), 12.
     [http://dx.doi.org/10.3390/c3020012]

[27]   dos Reis, G.S.; Adebayo, M.A.; Sampaio, C.H.; Lima, E.C.; Thue, P.S.; de Brum, I.A.S.; Dias, S.L.P.; Pavan, F.A. Removal of phenolic compounds from aqueous solutions using sludge-based activated carbons prepared by conventional heating and microwave-assisted pyrolysis. *Water Air Soil Pollut.,* **2017**, *228*(1), 33.
     [http://dx.doi.org/10.1007/s11270-016-3202-7]

[28]   Valili, S.; Siavalas, G.; Karapanagioti, H.K.; Manariotis, I.D.; Christanis, K. Phenanthrene removal from aqueous solutions using well-characterized, raw, chemically treated, and charred malt spent

rootlets, a food industry by-product. *J. Environ. Manage.,* **2013**, *128*, 252-258.
[http://dx.doi.org/10.1016/j.jenvman.2013.04.057] [PMID: 23764506]

[29]   Chen, B.; Chen, Z. Sorption of naphthalene and 1-naphthol by biochars of orange peels with different pyrolytic temperatures. *Chemosphere,* **2009**, *76*(1), 127-133.
[http://dx.doi.org/10.1016/j.chemosphere.2009.02.004] [PMID: 19282020]

[30]   Zhang, P.; Sun, H.; Min, L.; Ren, C. Biochars change the sorption and degradation of thiacloprid in soil: Insights into chemical and biological mechanisms. *Environ. Pollut.,* **2018**, *236*, 158-167.
[http://dx.doi.org/10.1016/j.envpol.2018.01.030] [PMID: 29414336]

[31]   Klasson, K.T.; Ledbetter, C.A.; Uchimiya, M.; Lima, I.M. Activated biochar removes 100 % dibromochloropropane from field well water. *Environ. Chem. Lett.,* **2013**, *11*(3), 271-275.
[http://dx.doi.org/10.1007/s10311-012-0398-7]

[32]   Yu, F.; Li, Y.; Han, S.; Ma, J. Adsorptive removal of antibiotics from aqueous solution using carbon materials. *Chemosphere,* **2016**, *153*, 365-385.
[http://dx.doi.org/10.1016/j.chemosphere.2016.03.083] [PMID: 27031800]

[33]   Zhou, Y.; Liu, X.; Xiang, Y.; Wang, P.; Zhang, J.; Zhang, F.; Wei, J.; Luo, L.; Lei, M.; Tang, L. Modification of biochar derived from sawdust and its application in removal of tetracycline and copper from aqueous solution: Adsorption mechanism and modelling. *Bioresour. Technol.,* **2017**, *245*(Pt A), 266-273.
[http://dx.doi.org/10.1016/j.biortech.2017.08.178] [PMID: 28892700]

[34]   Gwenzi, W.; Chaukura, N.; Noubactep, C.; Mukome, F.N.D. Biochar-based water treatment systems as a potential low-cost and sustainable technology for clean water provision. *J. Environ. Manage.,* **2017**, *197*, 732-749.
[http://dx.doi.org/10.1016/j.jenvman.2017.03.087] [PMID: 28454068]

[35]   Kaetzl, K.; Lübken, M.; Uzun, G.; Gehring, T.; Nettmann, E.; Stenchly, K.; Wichern, M. On-farm wastewater treatment using biochar from local agroresidues reduces pathogens from irrigation water for safer food production in developing countries. *Sci. Total Environ.,* **2019**, *682*, 601-610.
[http://dx.doi.org/10.1016/j.scitotenv.2019.05.142] [PMID: 31128373]

[36]   Viglašová, E.; Galamboš, M.; Danková, Z.; Krivosudský, L.; Lengauer, C.L.; Hood-Nowotny, R.; Soja, G.; Rompel, A.; Matík, M.; Briančin, J. Production, characterization and adsorption studies of bamboo-based biochar/montmorillonite composite for nitrate removal. *Waste Manag.,* **2018**, *79*, 385-394.
[http://dx.doi.org/10.1016/j.wasman.2018.08.005] [PMID: 30343768]

[37]   Yin, Q.; Liu, M.; Ren, H. Biochar produced from the co-pyrolysis of sewage sludge and walnut shell for ammonium and phosphate adsorption from water. *J. Environ. Manage.,* **2019**, *249*, 109410.
[http://dx.doi.org/10.1016/j.jenvman.2019.109410] [PMID: 31446122]

[38]   Ajmal, Z.; Muhmood, A.; Dong, R.; Wu, S. Probing the efficiency of magnetically modified biomass-derived biochar for effective phosphate removal. *J. Environ. Manage.,* **2020**, *253*, 109730.
[http://dx.doi.org/10.1016/j.jenvman.2019.109730] [PMID: 31665689]

[39]   Wang, P.; Ma, J.; Shi, F.; Ma, Y.; Wang, Z.; Zhao, X. Behaviors and effects of differing dimensional nanomaterials in water filtration membranes through the classical phase inversion process: A review. *Ind. Eng. Chem. Res.,* **2013**, *52*(31), 10355-10363.
[http://dx.doi.org/10.1021/ie303289k]

[40]   Jhaveri, J.H.; Murthy, Z.V.P. A comprehensive review on anti-fouling nanocomposite membranes for pressure driven membrane separation processes. *Desalination,* **2016**, *379*, 137-154.
[http://dx.doi.org/10.1016/j.desal.2015.11.009]

[41]   Wu, H.; Tang, B.; Wu, P. MWNTs/polyester thin film nanocomposite membrane: An approach to overcome the trade-off effect between permeability and selectivity. *J. Phys. Chem. C,* **2010**, *114*(39), 16395-16400.
[http://dx.doi.org/10.1021/jp107280m]

[42]　Brunet, L.; Lyon, D.Y.; Zodrow, K.; Rouch, J-C.; Caussat, B.; Serp, P.; Remigy, J-C.; Wiesner, M.R.; Alvarez, P.J.J. Properties of membranes containing semi-dispersed carbon nanotubes. *Environ. Eng. Sci.,* **2008**, *25*(4), 565-576.
[http://dx.doi.org/10.1089/ees.2007.0076]

[43]　Kim, S.; Jinschek, J.R.; Chen, H.; Sholl, D.S.; Marand, E. Scalable fabrication of carbon nanotube/polymer nanocomposite membranes for high flux gas transport. *Nano Lett.,* **2007**, *7*(9), 2806-2811.
[http://dx.doi.org/10.1021/nl071414u] [PMID: 17685662]

[44]　Dumée, L.; Sears, K.; Schü tz, J.; Finn, N.; Duke, M.; Gray, S. Carbon nanotube based composite membranes for water desalination by membrane distillation. *Desalination Water Treat.,* **2010**, *17*(1-3), 72-79.
[http://dx.doi.org/10.5004/dwt.2010.1701]

[45]　Maphutha, S.; Moothi, K.; Meyyappan, M.; Iyuke, S.E. A carbon nanotube-infused polysulfone membrane with polyvinyl alcohol layer for treating oil-containing waste water. *Sci. Rep.,* **2013**, *3*(1), 1509.
[http://dx.doi.org/10.1038/srep01509] [PMID: 23518875]

[46]　Chan, W.F.; Chen, H.; Surapathi, A.; Taylor, M.G.; Shao, X.; Marand, E.; Johnson, J.K. Zwitterion functionalized carbon nanotube/polyamide nanocomposite membranes for water desalination. *ACS Nano,* **2013**, *7*(6), 5308-5319.
[http://dx.doi.org/10.1021/nn4011494] [PMID: 23705642]

[47]　Zhao, X.; Cheng, L.; Jia, N.; Wang, R.; Liu, L.; Gao, C. Polyphenol-metal manipulated nanohybridization of CNT membranes with FeOOH nanorods for high-flux, antifouling and self-cleaning oil/water separation. *J. Membr. Sci.,* **2020**, *600*, 117857.
[http://dx.doi.org/10.1016/j.memsci.2020.117857]

[48]　Wen, Y.; Yuan, J.; Ma, X.; Wang, S.; Liu, Y. Polymeric nanocomposite membranes for water treatment: A review. *Environ. Chem. Lett.,* **2019**, *17*(4), 1539-1551.
[http://dx.doi.org/10.1007/s10311-019-00895-9]

SUBJECT INDEX

A

Absorption 23, 36
  electromagnetic 23
Acid(s) 3, 19, 20, 23, 44, 45, 46, 48, 65, 82,
    83, 100, 129, 143
  hydrolysis 3
  lactic 19, 44, 45, 46, 48
  lauric-stearic (LSA) 83
  nitric 45
  palmitic (PA) 23, 82, 83, 100
  phosphoric 65, 143
  polylactic 19, 20, 46, 129
Acidic soils 116
Acrylonitrile butadiene styrene (ABS) 49, 100
Additive manufacturing technologies 87
Adsorbents, natural 35, 51
Adsorption 21, 63, 141, 142, 143
  dye 21
  fuchsine 63
Agglomeration effects 94, 141
Agricultural waste disposal 72
Agriculture 36, 72, 73
  industry 73
  sector 36, 72
Alloys, reinforced 116
Almond business 75
Almond shell biochar 78, 79, 80, 81, 82, 144
  activated 81
  powder 78
Antibacterial agents 145
Antibiotic toxicity 144
Anticorrosion properties 25
Antimicrobial 16, 25, 26
  agents 26
Applications, industrial 142
Automotive fields 42

B

Balanced mechanical properties 47
Ball milling process 92, 94

Behavior 3, 15, 25, 39, 40, 42, 49, 122, 131,
    134
  crystallization 40
  hollow nano-fiber 15
  pliable processing 25
  shear-thinning 49
  thermal 131, 134
Bio-sourced polyamides 44
Biocarbon 1, 24, 25, 95
  fibers 24
  nanoparticles 1, 25
  polymer nanocomposite materials 95
Biochar 36, 38, 39, 41, 42, 44, 45, 46, 48, 49,
    51, 52, 74, 79, 80, 83, 93, 115, 117, 123,
    129, 142, 143, 144
  biomass-derived 74, 80
  carbon-reinforced polymer composites 129
  nanomaterial-engineered 83
  properties 79
  sludge-based 142
  thermoplastic Polymer Composites of
    biochar 45
  UHMWPE composites 41, 42
Biocomposite materials 98, 104
  polymeric 98
  sustainable 98
Biocomposites 20, 24, 98, 99, 100, 101, 102,
    105, 106, 107, 108, 110, 111
  polymeric 99
  wood-derived biochar-reinforced
    polypropylene 24
Biodegradability 20, 100
Biodegradable plastic 25, 94
Biomass 2, 3, 6, 24, 35, 60, 61, 72, 73, 74, 83,
    84, 89, 127
  almond 73, 74
  animal-derived 2
  bamboo 83
  carbon 24
  disposal 72
  lignocellulosic 6
  natural 89

www.ingramcontent.com/pod-product-compliance
Lightning Source LLC
Chambersburg PA
CBHW041708210326
41598CB00007B/572